Statistical Approaches to Gene × Environment Interactions for Complex Phenotypes

Statistical Approaches to Gene × Environment Interactions for Complex Phenotypes

edited by Michael Windle

The MIT Press
Cambridge, Massachusetts
London, England

This book was set in Stone Serif Std by Toppan Best-set Premedia Limited. Printed and bound in the United States of America.

Library of Congress Cataloging-in-Publication Data

Names: Windle, Michael T., editor.
Title: Statistical approaches to gene × environment interactions for complex phenotypes / Windle, Michael, ed.
Description: Cambridge, MA : The MIT Press, 2016. | "This edited volume stemmed from the 3rd Annual Symposium of the University of Georgia's Center for Contextual Genetics and Prevention Science (CGAPS) ... was held on June 19–20, 2012 in Athens, GA"—Preface. | Includes bibliographical references and index.
Identifiers: LCCN 2015041250 | ISBN 9780262034685 (hardcover : alk. paper)
Subjects: LCSH: Genotype-environment interaction—Statistical methods.
Classification: LCC QH438.5 .S73 2016 | DDC 576.8/5—dc23
LC record available at http://lccn.loc.gov/2015041250

10 9 8 7 6 5 4 3 2 1

Contents

Preface

This edited volume stemmed from the 3rd Annual Symposium of the University of Georgia's Center for Contextual Genetics and Prevention Science (CGAPS). With the theme "Methodological and Statistical Issues in Gene × Environment Research," this exciting symposium was held on June 19–20, 2012, in Athens, Georgia. Scientific and technological advances in genomics and findings from the Human Genome Project, as well as subsequent Genome-Wide Association (GWA) studies, have spawned stimulating discussions about the role(s) of genes for a broad range of diseases and traits and reconceptualizations and reformulations of basic models and associated assumptions. For instance, before the completion of much of the research from the Human Genome Project and GWA studies, the common-disease common-variant model (CDCV) served as a principal orientation to many studies in genomics. According to this model, the more intensive and extensive gene searches afforded by advanced sequencing technology on a larger number of subjects would yield the identification of one or a few genetic loci (or variants) that would be associated with a given disease; these identified genes could then be targeted in research to further ensure their status as causal variants that if (or when) confirmed would then serve as the foundation for biomedical treatments (e.g., pharmaceutical products, vaccines). Unfortunately, as often occurs, nature is not nearly as obliging as our models would assume, and subsequent research has recognized the limits of the CDCV model. Based on research findings, it has become clear that other sources of genomic data (e.g., rare variants, copy number variation) are highly relevant, as are contributions from the environment (e.g., exposure to stressful events or to pesticides or chemical agents).

In this volume, the focus is specifically on the potential role(s) that gene by environment interactions (G × E) may have on a range of complex diseases and traits. By design, an interdisciplinary group of highly accomplished investigators was invited to present at the symposium. Like many areas of science, it is easy to become discipline-bound and to overlook the conceptual and empirical contributions of colleagues in cognate disciplines who are studying similar, though not identical, research issues but perhaps for different phenotypes. By assembling and facilitating exchange among scholars from different disciplines (e.g., biostatistics, statistical genetics, clinical and developmental psychology) we sought to broaden the lens of investigator exposure that nevertheless centered on the common interest of G × E interactions. Consistent with the integrative and synergistic nature of CGAPS we thought that it was especially beneficial to bring together scholars who are investigating G × E interactions from various backgrounds in medical and psychiatric genetics with scholars from various backgrounds in the behavioral sciences and public health.

We were very fortunate to have this symposium cosponsored at the University of Georgia by Vice President for Research David C. Lee of The Center for Family Research and The Owens Institute for Behavioral Research. Without their generous support, the symposium would not have come to fruition. We would also like to thank the National Institute on Drug Abuse for their support of CGAPS (P30DA027827) and our competing renewal application that revised our title to the Center for Translational & Prevention Science (P30DA027827) (www.ctaps.uga.edu). Finally, I would like to express my appreciation to the National Institute of Alcohol Abuse and Alcoholism for their continued support of my Career Research Scientist Award (K05AA021143) that further provided me the opportunity to pursue this volume. It is my hope that this volume stimulates further thinking about conceptualizing, measuring, and translating G × E interactions via an inclusive, open forum of intellectual exchange and discussion. By utilizing the joint contributions of the many contributors to the study of G × E interactions, national and international goals for public health may be optimized.

1 Introduction and Overview of Statistical Approaches to Gene × Environment Interactions for Complex Phenotypes

Michael Windle

High-throughput sequencing has spawned a new era of inquiry with regard to fundamental questions about the discipline of genomics and long-standing underlying assumptions in the field (e.g., the number of protein coding genes, the role of junk DNA, and the common disease-common variant model [CDCV]). Prior "thinking" in genomics that guided the highly influential Human Genome Project focused almost exclusively on the approximately 2 percent of DNA that is involved in the sequencing of proteins that are critical to the development of cells and the subsequent development of bodily organs and other biological attributes (e.g., neurons, neurotransmitters, hormones). This orientation assumed that by identifying genes associated with these sequenced data major insights would be provided about the locus or loci associated with diseases that would then provide targets for various medical interventions (e.g., pharmaceutical products, gene therapies, vaccines). This cogent picture of how the genome, based on findings from the Human Genome Project, would provide the roadmap to guide translational research and personalized medicine did not materialize. Rather, findings from the Human Genome Project strongly indicated a much more complex pattern that prohibited the rapid translation of this genomic information to multiple diseases in the manner initially anticipated. The outgrowth has been a flurry of research adjusting to a reconceptualization that recognizes that for non-Mendelian disorders—hereafter referred to as complex phenotypes because of reference not only to diseases (disorders) but also to traits—there are many sources of genomic variation that are influential beyond common variants.

Some of these other sources of influence on complex phenotypes include the role of multiple, small effect size, gene influences (e.g., thousands of single nucleotide polymorphisms; SNPs) rather than one or few SNPs, low frequency (rare) alleles that may be of larger magnitude than common variants, structural variation such as copy number variation or copy neutral variation (e.g., translocations and inversions), variation in noncoding regions that may exert regulatory control on transcriptional processes, and epigenetic influences that contribute to transcriptional efficiency. Francis Collins, Director of the National Institutes of Health (NIH), and Harold Varmus, Director of the National Cancer Institute, have also proposed that the integration of data across the multiple-omics fields such as genomics, proteomics, and metabolomics may be necessary to more fully capture underlying biological causes of disease and contribute to precision medicine [1].

Yet another possibility for understanding complex phenotypes is the role of gene × environment (G × E) and G × G interactions. A major impetus for the current book stemmed largely from at least three sources. First, large-scale Genome-Wide Association (GWA) studies of complex phenotypes have typically yielded low magnitude effects (i.e., amounts of variance accounted for in the phenotype by the genes). For example, with more than 250,000 cases, optimal gene search strategies for height yielded estimates of less than 5 percent of the variance accounted for in height due to genetic influences [2]. This finding was not unique, since effect size (ES) estimates have rarely exceed 10 percent and are typically less than 5 percent for behavioral traits and psychiatric disorders [3, 4]. Furthermore, in many instances the biological plausibility of the identified statistically significant SNPs are not apparent, in large part because the selection of genes is driven by statistical algorithms in GWA studies, with biological plausibility a desirable, but definitely not assured, product of the analyses. The obtained ES estimates in GWA studies using very large sample sizes dwarf expectations based on heritability studies that commonly report heritability estimates in the range of 40–80 percent contingent on the phenotype. This "missing heritability" issue has increased sensitivity to other possible sources of variation that may account for complex phenotypes, including G × E interactions. Second, given the current high-throughput sequencing technology and the ability to generate truly "Big Data," issues have arisen surrounding the burden of multiple hypothesis testing (i.e., statistical corrections or adjustments to nominal alpha levels due to multiple testing). For example, it is common in GWA studies to use a stringent alpha level such as $p < 5 \times 10^{-8}$ as the threshold to determine statistical significance. This is an important issue for GWAS and Next Generation Sequencing (NGS) applications, and it is compounded for the investigation of many (most) complex phenotypes where the collection of more than 200,000 cases is highly unlikely due to exorbitant costs (e.g., studies in neuroimaging or randomized clinical trials).

Third, for many complex phenotypes (e.g., depression, diabetes, obesity, substance use), there is substantial evidence that while genetic influences are important, so are environmental influences; moreover, there is substantial evidence from both behavior genetic studies (e.g., twin and adoptee studies) and molecular genetic studies (both human and infrahuman) that genes commonly interact with environmental factors in predicting complex phenotypes. Many years ago, Anastasi [5] described the nature-nurture debate as one that should focus on *how* genes and environments interact to produce phenotypes and not on how much to attribute to each component. Given the current state of the field regarding complex phenotypes and our ability to measure both individual genes and environmental factors, this argument may be beneficially revisited within a broader, developmental multilevel approach.

This book is organized into eleven chapters, including the current chapter, which introduces some major G × E issues of relevance to this volume and a brief overview of each chapter. Chapters 2–6 delve into the issue of the burden of multiple hypothesis testing and data reduction (e.g., via screening) by providing different statistical approaches or strategies to address G × E and G × G interactions with high-throughput sequenced data such as that provided in GWA and NGS studies. More specifically, in chapter 2, Kooperberg, Dai, and Hsu provide the history and rationale for the development of two-stage procedures to identify G × E and G × G interactions. The first (exploratory) stage involves screening SNPs for their potential to be involved in interactions and the second (testing/confirmation) stage involves testing only those SNPs that passed the initial screening. This reduction in the number of SNPs tested in the second stage can increase the power of detecting G × E and G × G interactions by reducing the burden of multiple comparison corrections. This reduction is important because it takes substantially more subjects (e.g., up to four times as many) to test for a G × E interaction that is equally powerful as a test for a G main effect. Kooperberg et al. provide discussions of the use of this two-stage approach with randomized control trials, additive (as well as multiplicative) interactions, G × G interactions, and G × E interactions for quantitative traits.

In chapter 3, Tzeng and Maity propose that the analytic focus should shift from individual polymorphisms (e.g., SNPs) to the gene level, and describe marker-set approaches to identifying and evaluating G × E and G × G interactions. The authors suggest that focusing on the gene level rather than individual SNPs is advantageous for several reasons, including that SNPs within a gene tend to function in a coordinated fashion, gene-level analysis incorporates linkage disequilibrium (LD) information from all SNPs of a gene simultaneously, and the joint analysis of SNPs at the gene level can be more powerful statistically than individual SNP analysis. Tzeng and Maity describe both fixed and random effects models for marker-set based approaches, and they discuss both strengths and limitations of these alternative modeling procedures for assessing G × E and G × G interactions.

In chapter 4, Jiao presents set-based approaches that focus on identifying G × E interactions rather than set-based approaches that are based primarily on detecting G main effects (e.g., via marginal effects). Jiao reviews both his own research and the development of his Set Based gene EnviRonment InterAction test (SBERIA), as well as another set-based G × E approach referred to as GESAT. GESAT extends the variance component test of the SNP-set Kernel Association Test (SKAT) to evaluate G × E effects while incorporating the main SNP effects as covariates. While both of these approaches (SBERIA and GESAT) have outperformed other benchmark methods (e.g., likelihood ratio test) and have been demonstrated to retain the appropriate type 1 error rate, in this chapter Jiao conducts simulation studies to compare findings for SBERIA and GESAT approaches and identifies associated strengths and limitations of the respective methods.

Wang, in chapter 5, also proposes the use of a gene-based approach to study G × E and G × G interactions via the use of a partial-least square (PLS) approach. The focus on the gene level reduces the multiple testing burden because the number of tests is based on gene-sets rather than individual SNP markers. The PLS approach involves extracting a latent component of multiple SNPs within a gene region that maximizes the correlation between the set of SNPs, the environmental factor, and the disease. The dimension reduction of multiple SNPs is influenced both by the marginal effects of SNPs and linkage disequilibrium (LD) among the SNPs in the sample. The

PLS approach can be used with a genotype matrix of a selected gene or with an interaction matrix of two genes or one gene and an environmental factor. Wang suggests that the PLS approach may reduce the dimension of interactions to be tested in an efficient and effective way without loss of useful signal information.

In chapter 6, Todorov provides a description and overview of the RELIEF algorithm and its extensions to identify and weight relevant features of a trait of interest from irrelevant features. RELIEF stems from a machine-based learning approach and is based on a nonparametric statistical method for feature selection rather than the more commonly used parametric statistical methods in the study gene and $G \times E$ relationships. The feature selection algorithm seeks to screen or reduce relevant features from a large number of attributes, such as a few hundred thousand SNPs in genomics research. By acting as a statistical filter, RELIEF seeks to identify relevant subsets of genes, $G \times G$, and $G \times E$ interactions in a computationally tractable manner, no small challenge given the abundance of data available via high-throughput sequencing technology. Todorov describes some of the important statistical issues surrounding the use of the RELIEF algorithm (and its extensions), such as initially selecting nearest neighbors in analyses that fundamentally rely on pairwise differences among the multiple elements or features of the trait (e.g., SNPs), alternative weighting approaches for the identified neighbors, and iterative versions of RELIEF. He also provides some simulation findings for evaluating gene and environment factors using the RELIEF algorithm, with findings suggesting that this approach outperformed more standard logistic regression screening approaches for the low-powered simulation design. Additional research on statistical properties of RELIEF is proposed to further examine its utility for applications in the study of $G \times E$ and $G \times G$ interactions.

Each of these five chapters (chapters 2–6) focuses on methods, models, or strategies that address salient conceptual and statistical issues in the investigation of $G \times E$ and $G \times G$ interactions. These include methods that attempt to address issues related to the multiple testing burden, improve screening and testing procedures to more effectively identify SNPs or gene-level analyses, and improve specific statistical tests, estimation procedures, and analytical approaches. These five chapters may be described as more discovery-oriented with regard to identifying and pursuing $G \times E$ and $G \times G$ interactions. By comparison, chapters 7–11 focus on specific complex phenotypes (e.g., obesity, substance use), research designs (e.g., randomized control trials), or combined methods (e.g., SNP and gene expression) that may advance the study of G × E interactions.

In chapter 7, McCaffery and Doyle present evidence on the value of using randomized clinical trials (RCTs) to investigate G × E interactions in obesity research. In GWA studies, a large number of typically low-effect-size SNPs have been associated with BMI and obesity. One of the identified SNPs is FTO, a genetic variant that is associated with feeding behaviors such as preference for energy dense food, greater consumption of fat and calories, and reduced satiety. FTO has significantly interacted with a number of environmental characteristics (e.g., physical activity, fried food consumption) in predicting BMI and obesity. Building upon these findings, RCTs have been conducted to see if (or how) genetic loci interact with treatment in producing weight loss. McCaffery and Doyle describe the strengths of RCTs to evaluate G × E interactions given that the environment (i.e., treatment vs. control condition) is randomly assigned. The randomization of the environment has a host of benefits for inferential statistics and reduces if not eliminates concerns about G × E correlations (i.e., individuals with certain characteristics, including genes, selecting into specific environments). The chapter provides some rich illustrations taken from the larger obesity intervention research literature, including the authors' own studies, to demonstrate how identified genetic factors such as FTO may interact with targeted features of treatment (e.g., physical activity, dietary modifications) to impact weight loss, thereby linking G × E research to potential clinical practice in public health.

In chapter 8, Mukherjee, Chen, Ko, He, Lee, Zhang, and Park highlight the importance of longitudinal research designs and statistical models to investigate G × E relationships. This chapter provides a history of alternative statistical models that have been used to study G × E interactions with longitudinal data, as well as a critique of their relative strengths and weaknesses. Of importance, the models reviewed incorporate information on both time invariant and time varying exposures to characterize more dynamic patterns of change across time. The authors use longitudinal data from the Normative Aging Study to illustrate some of the longitudinal models by focusing on pulse pressure, which is a risk factor for arterial stiffness. A reduced set of SNPs that have been identified in this substantive area of research were identified and the outcome measure was level of lead in the tibia bone. They describe their approach as a pathway orientation toward identifying how environmental exposures across time may influence changes in bone lead levels. This illustration provides clarity on issues raised in modeling G × E longitudinal data that generalize to other areas of health and to other phenotypes. This chapter also raises critical questions about where we have been and where we need to go in the modeling of G × E interactions with longitudinal data.

In chapter 9, Ahmed and Kinnally provide some interesting longitudinal examples and illustrations of how G × E influences may be studied with regard to neurobehavioral (brain) development in human and nonhuman primates. The chapter provides keen insight into two significant conceptual and methodological issues in the study of G × E interactions. First, is the importance of considering the findings from both human and nonhuman studies on genes and environment, thereby suggesting a more integrative lens for thinking about, planning, and interpreting research findings in G × E research. Second, they propose the use of multiple methods to investigate G × E interactions, including in their applications the use of both SNP-based and micro-array-based methods. With the quite massive increases in available large data sources (e.g., genomics, proteomics, metabolomics), there will be clear benefits in future research to incorporate different methods or sources of data toward identifying underlying biological mechanisms. Furthermore, the use of longitudinal research designs to study G × E interactions on time-ordered change phenomena such as neurobehavioral development provides a promising approach to identify and translate basic research findings into practice.

In chapter 10, Windle explores the potential value of developing and using polygenic scores to investigate G × E interactions for complex phenotypes. That is, rather than using only one loci (e.g., one SNP or one VNTR), the idea is to develop multiple loci or polymorphisms that can be used as an aggregate score (or index) to be associated with phenotypes. If reliable and valid aggregated scores could be developed for specific phenotypes, then the burden of multiple testing would be reduced to the polygenic score(s) rather than each individual SNP score. Furthermore, to the extent that the polygenic score reflects the summation of multiple genetically informative SNPs, it is possible that effect size estimates for the phenotypes of interest will be higher. Literature relevant to measurement issues in forming aggregated polygenic indexes are provided in this chapter, as is a path-analytic-oriented conceptual approach to modeling G × E interactions using both aggregated and disaggregated polygenic indexes to measure predictors of substance use and substance disorders with both cross-sectional and longitudinal research designs.

In chapter 11, Thomas focuses on epigenetics and statistical models and simulations that would facilitate the investigation of epigenetics as a mediator of exposure-response relationships. Epigenetics refers to factors (influences) that alter gene expression without modifying the underlying DNA sequence. DNA methylation or other markers (e.g., posttranslational histone modifications) are used to identify epigenetic changes. For example, Fraga et al. [6] reported on findings using monozygotic twins who are genetically identical and had highly similar DNA methylation and histone acetylation patterns early in life; however, across time (age) these MZ twins had quite different methylation and histone acetylation patterns. These changes were possibly attributable to variable responses to different exposure histories, though they could also be due to random drift. There has also been support, primarily in the animal studies literature, for transgenerational changes in epigenetic effects, though quite limited research on this topic has been conducted with humans [7, 8]. Thomas provides research methods, analytical approaches, and simulations to test both individual and transgenerational changes in humans to further probe the possibility of identifying epigenetic effects and associated environmental exposures within a general statistical framework on mediation of exposure-response relationships through epigenetic effects.

This book is neither intended to be a handbook or comprehensive volume on G × E statistical issues and models, nor an exhaustive presentation of conceptual approaches related to G × E interactions; rather, it is designed to provide a representation both of statistical modeling and methodology-based approaches to investigating G × E interactions for complex phenotypes. Because most courses in basic genetics and behavioral epidemiology typically devote a limited amount of time to G × E interactions, this book could be used pedagogically as a supplement to provide a broader perspective on these interactions, or serve as a core reference for a molecular genetics or genetic epidemiology seminar devoted the study of G × E interactions of complex phenotypes. It could also be of benefit to many public health and behavioral scientists who are relatively new to field of G × E interactions to gain a rapid sense of many of the salient issues in the field, as well as some of the proposed solutions to these issues. The fields of genomics and other omics (e.g., proteomics, metabolomics) provide exciting opportunities to advance science and foster the goals of public health and a more individualized intervention approach (e.g., personalized medicine; precision medicine). The goals of these more individualized approaches would benefit greatly not only by advances in genomics and other omics, but also by incorporating information both on environments and their interactions with genomic and other biological material and regulatory processes (e.g., environmental signals to biological pathway responses). Such findings would thereby offer more flexible guidance to a broader range of prevention, intervention, and treatment targets and facilitate more tailored programs based on a fuller complement of G and E influences.

Acknowledgments

This research was supported by NIDA Grant Number DA027827 and NIAAA Grant Number K05AA021143. The contents are solely the responsibility of the authors and do not necessarily represent the official views of the National Institutes of Health.

References

1. Collins, F. S., & Varmus, H. (2015). A new initiative on precision medicine. *New England Journal of Medicine, 372*, 793–795.

2. Speliotes, E. K., Willer, C. J., Berndt, S. I., Monda, K. L., Thorleifsson, G., Jackson, A. U., et al. (2010). Association analyses of 249,796 individuals reveal 18 new loci associated with body mass index. *Nature Genetics, 42*, 937–948.

3. Manolio, T. A., Collins, F. S., Cox, N. J., Goldstein, D. B., Hindorff, L. A., Hunter, D. J., et al. (2009). Finding the missing heritability of complex diseases. *Nature, 461*, 747–753.

4. Plomin, R. (2013). Child development and molecular genetics: 14 years later. *Child Development, 84*, 104–120.

5. Anastasi, A. (1958). Heredity, environment, and the question "how?" *Psychological Review, 65*, 197–208.

6. Fraga, M. F., Ballestar, E., Paz, M. F., Ropero, S., Setien, F., Ballestar, M. L., et al. (2005). Epigenetic differences arise during the lifetime of monozygotic twins. *PANAS, 102*, 10604–10609.

7. McGowan, P. O., & Roth, T. L. (2015). Epigenetic pathways through which experiences become linked with biology. *Development and Psychopathology, 27,* 637–648.

8. Meaney, M. J. (2001). Maternal care, gene expression, and the transmission of individual differences in stress reactivity across generations. *Annual Review of Neuroscience, 24,* 1161–1192.

2 Two-Stage Procedures for the Identification of Gene × Environment and Gene × Gene Interactions in Genome-Wide Association Studies

Charles Kooperberg, James Y. Dai, and Li Hsu

Identification of gene × environment (G × E) interactions is often of great interest. Such interactions provide insight into disease etiology and underlying biological mechanisms. These interactions can also identify subgroups of subjects defined by environmental variables for which particular genetic variants have especially strong effects, or subjects with particular genotypes for which environmental variables have differential effects. In addition, they can potentially explain the missing heritability that has not been accounted for by common genetic variants.

Testing for (multiplicative) interactions can be straightforward. For example, the logistic regression model with a multiplicative gene-environment interaction is

$$\mathrm{logit}\,(Y = 1 \,|\, G, E) \;=\; \beta_0 + \beta_1 G + \beta_2 E + \beta_3 GE \,, \tag{2.1}$$

where Y is disease status, G is a single nucleotide polymorphism (SNP) that is often coded by the number of variant alleles, and E an environmental variable; we can easily add other variables X to the model. In this model we can test H_o: $\beta_3 = 0$ versus H_A: $\beta_3 \neq 0$ by fitting a standard generalized linear model. Note, however, that in model (2.1) we need approximately four times as many subjects for an interaction test that is equally powerful as a main effect test for β_1 [1]. In some scenarios other models may be more powerful, e.g., if G and E are known to be independent in the population and the disease (Y) is rare, testing H_o: $\alpha_1 = 0$ in a case-only model

$$\mathrm{logit}(E = 1 \,|\, G, Y = 1) \;=\; \alpha_0 + \alpha_1 G \,, \tag{2.2}$$

is considerably more powerful than the interaction test for β_3 in (2.1). In some other situations, different forms of the regression model may also lead to more powerful tests.

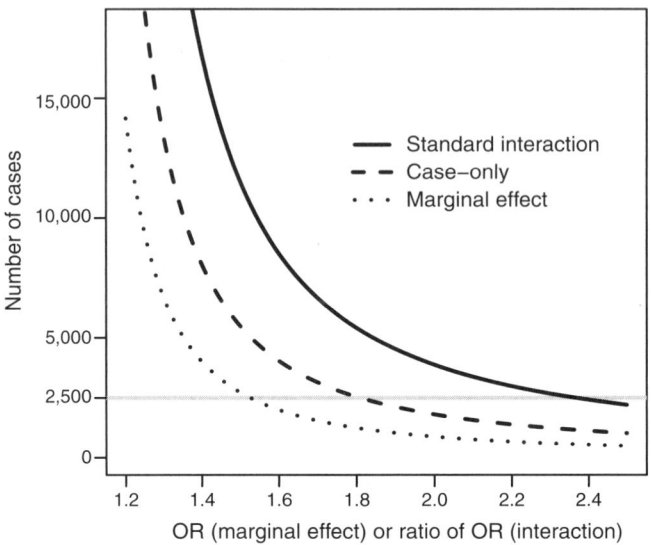

Figure 2.1
Sample size needed to identify an effect for a SNP with minor allele frequency 0.2 and a binary environmental variable with $P(E = 1) = 0.5$ in a case-control study with the same number of cases and controls and a Bonferroni correction for 1,000,000 SNPs.

In figure 2.1 we show the required sample size to identify a SNP with minor allele frequency 0.2 as having a main effect that is significantly different from 0 at a genome-wide significance level of $\alpha = 5 \times 10^{-8}$ (a Bonferroni correction for 1 million SNPs) with a power of 80 percent, as well as the required sample sizes for identifying an interaction between a gene and an independent binary environmental factor with frequency 0.5, using a case-only analysis H_o: $\alpha_1 = 0$ in (2.2) or a standard case-control analysis H_o: $\beta_3 = 0$ in (2.1). Recognizing that most genome-wide association studies (GWAS) are only adequately powered to identify main effects, it is clear that these studies would be severely underpowered for identifying interactions. For example, for the setup of figure 2.1 a study with 2,500 subjects (gray line) for each group, would have power for a main effect of approximately 1.53 (corresponding to $\beta_1 = 0.42$), an interaction effect using a case-only analysis of approximately 1.81 ($\alpha_1 = 0.59$), and an interaction effect using a case-control analysis of approximately 2.37 ($\beta_3 = 0.86$). This figure makes it very clear that standard analyses will typically not have sufficient power to identify G × E effects.

If we knew which of the many SNPs in a GWAS were most likely to be involved in an interaction, we could restrict ourselves to testing only those SNPs, reduce the multiple-comparisons correction, and thus improve the power of identifying interactions substantially. In this chapter, we discuss two main test statistics and their combinations for identifying such SNPs: (1) Select those SNPs that show some marginal effect. The motivation of this strategy is that if there were an interaction, we should see a genetic effect in at least one of the environmental groups, which results in an overall marginal effect, unless we are in the unfortunate situation that genetic effects in different environmental groups exactly cancel each other out. (2) Select those genes that show a correlation between G and E in all samples of cases and controls combined. The intuition behind this strategy is that if there is an interaction effect on Y, the correlation between G and E differs between cases ($Y = 1$) and controls ($Y = 0$). Thus, in at least one response stratum the correlation between G and E is not zero. With a similar reasoning as for screening on the main effects, if there is a difference of correlation between cases and controls, we "hope" that the correlations in the cases and controls do not exactly cancel. After either of these screening procedures, our goal would be to identify a subset of SNPs for G × E interaction and correct only for the number of SNPs we test for interaction, rather than exhaust the number we could have tested. To do this we need to carefully consider whether such a procedure maintains the overall type 1 error.

In the next section, "Methods," we first consider the advantages situations when the type 1 error rate of a two-stage procedure is and is not maintained. Then we discuss screening strategies and how they can be combined to facilitate testing of G × E interactions. The strategies involve both the tests used at the screening stage and at the testing of interaction stage, as well as the combination of significance levels. In "Practical Issues," we discuss six topics in more detail:

• When does a two-stage strategy help and how does the stage one screening influence the results?
• How does linkage disequilibrium (LD, correlation between the SNPs) influence these procedures?
• How can we exploit gene-environment independence in randomized clinical trials?

- Do these approaches work with models for additive interactions?
- How are these methods applicable to gene × gene interactions?
- How can we use these methods with quantitative traits?

Methods

Theoretical Framework of Two-Stage Testing

A two-stage testing procedure was developed by Dai et al. [2], where in the first stage a hypothesis K_{oj}: $\zeta_j = 0$ is tested at significance level $0 < \alpha_0 < 1$ for each SNP j, where we reject m_0 of the m hypothesis (so that $m_0/m \to \alpha_0$), and at the second stage we test G × E using a test H_{oj}: $\theta_j = 0$ at level $\alpha/(2m_0)$, then

$$\lim_{m\to\infty}\lim_{n\to\infty}P\left(\text{reject both } K_{oj} \text{ and } H_{oj} \text{ for at least one } j\right) \le \alpha,$$

where n is the sample size, provided

$$\text{cov}\left\{n^{1/2}\left(\hat{\zeta}_j - \zeta_j\right), n^{1/2}(\hat{\theta}_j - \theta_j)\right\} \to 0, \text{ for all } j \in \{1,...,m\}.$$

The interpretation of this condition is simple: if the screening test and the test for G × E are (asymptotically) independent we can justify controlling only for the number of G × E tests that we conduct, rather than for the number we could have conducted. The independence of the two estimators is conditional on the true values (ζ_j and θ_j). Across different SNPs j, ζ_j can be informative toward θ_j, i.e., a non-zero θ_j may indicate a non-zero ζ_j, thereby forming a basis for potentially improving power of a G × E test by first screening for ζ_j.

Dai et al. [2] established that for generalized linear models, as well as the Cox proportional hazards models, a screening test statistic for the marginal association, as discussed in what follows, is independent of the test statistic for the usual (multiplicative) interaction (e.g., (2.1)), as well as of the case-only estimator (2.2). Here, we also establish that while screening using the correlation between G and E is independent of the test statistic for the usual (multiplicative) interaction, it is not independent of the case-only estimator. This clearly shows that using a two-stage procedure involves choosing a pair of tests. Not every screening test statistic can be matched up with every test for G × E.

Screening Statistics

Marginal Association of G with Disease Risk This series of research studies began when Kooperberg and LeBlanc [3] proposed to test for gene × gene (G × G) interactions by first testing all SNPs G_j in the regression model

$$\text{logit}\left(Y = 1 \,|\, G_j\right) \;=\; \gamma_0 + \gamma_1 G_j, \tag{2.3}$$

at an initial significance level α_0 and then testing only those interactions $G_j G_k$ for which both SNPs j and k are significant in (2.3), using a significance level that corrects for the number of interaction tests conducted. As Millstein [4] notes, the results from Dai et al. [2] for G × E interactions also cover G × G procedures like this one. The generalization to G × E is immediate, as discussed in Kooperberg et al. [5].

Correlation of G and E Murcray et al. [6] proposed to identify G × E by first identifying SNPs G_j that are marginally correlated with a binary environmental factor E in the model

$$\text{logit}\left(E = 1 \,|\, G_j\right) \;=\; \delta_0 + \delta_1 G_j, \tag{2.4}$$

which is effectively the same as testing in (2.2), but for cases and controls combined. SNPs that are significant at an initial significance level α_0 are then tested in the standard logistic regression model (2.1). The paper by Murcray et al. [6] seems partly motivated by Millstein et al. [7], who had proposed screening for marginal correlation between SNPs followed by a test for interaction. In that paper, however, the goal was to identify SNPs that were marginally correlated with the response. Lewinger et al. [8] proposed screening on SNP-SNP correlations to identify potentially relevant G × G interactions.

Illustration of Why These Two Screening Statistics Are Useful in Informing G × E In the introduction we provided some reasoning as to why screening for marginal effects is useful: Unless the interaction is purely qualitative, that is, the effect of the SNP in one stratum exactly cancels the effect in the other stratum, an interaction effect will result in a main effect when marginalizing over the environmental factor, suggesting that a screen on the marginal effect is useful. This would especially be the case if there is only a genetic effect for, say, high values of E, and for low values of E there is no effect. If the genetic effects are in opposite directions for different E, the marginal screen would be less effective (and could conceivably lead to a situation where a two-stage procedure with a marginal screen loses power).

The argument of why screening for G-E correlation helps is similar: If there is a G × E, the correlation between G and E is different for different values of the response; thus, unless correlations exactly cancel, there is a non-zero G-E correlation over all subjects, suggesting that filtering on G-E correlation is useful. As an illustration, we simulated a data set of 500 cases and 500 controls. We generated one binary E with $P(E = 1) = 0.5$ and 200 (binary) SNPs each with $P(G = 1) = 0.5$; the main effects of each of these SNPs are generated as null. We randomly selected twenty SNPs to have an interaction effect on disease risk with interaction odds ratios coming from a Uniform[1.5,3] distribution (thus, β is between 0.4 and 1.1). For each SNP we calculated the correlation between G and E, the marginal association of G with disease risk, the interaction effect of G × E with disease risk using the conventional logistic regression estimator and the interaction effect of G × E using the case-only estimator. Figure 2.2 shows the pairwise scatterplots of these associations of 200 SNPs with the x's indicating the SNPs that have interaction effect and the o's indicating the SNPs with no interaction.

We make several observations based on this figure. First, SNPs with interaction effects also have larger values for the G-E correlation and the marginal association of G with disease risk, indicating that these two screening statistics can indeed help prioritize SNPs for G × E testing. Second, the case-control G × E estimator is independent of both the marginal association estimator screening and G-E correlation estimator when the interaction effect is 0, since the o's show no relation. Third, the case-only G × E estimator remains independent of the marginal association estimator; however, there is a clear linear relationship between the case-only G × E estimator and the G-E correlation, suggesting that if one performed G-E correlation screening followed by the case-only estimator, the type 1 error would be inflated.

A simple simulation, in a situation with gene-environment independence, which was shown in Dai et al. [2], is repeated as figure 2.3. Approximately 1,000 cases and 1,000 controls were sampled to test gene-environment interaction with effect size log(2) among 10,000 SNPs. We varied the main genetic effect to generate either qualitative or quantitative interaction. We see from this figure that depending on the situation different procedures can be the most powerful: When the disease is rare and there

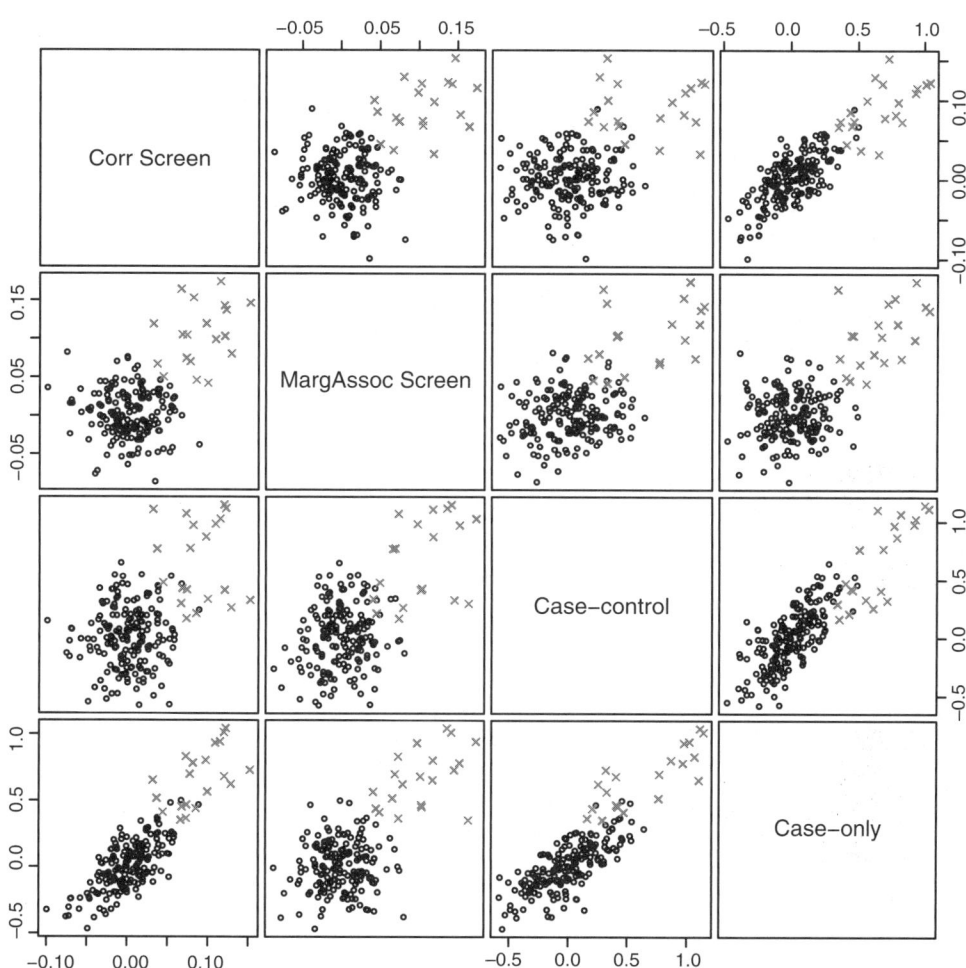

Figure 2.2

Estimates of regression coefficients and correlations for 200 SNPs. The x's are SNPs that have an interaction effect. Corr Screen refers to the correlation between G and E; MargAssoc Screen refers to the association of G with disease risk; Case-control and Case-only refer to the interaction effects of G × E with disease risk by models (2.1) and (2.2), respectively.

(a)

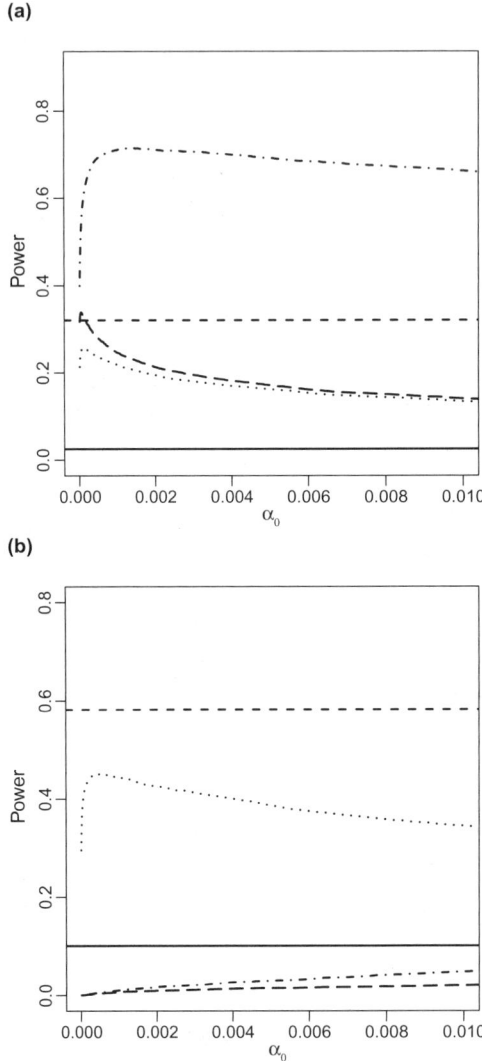

(b)

Figure 2.3

Power comparison of the two-stage procedures for detecting gene-environment interaction. The four panels represent the scenarios (a) rare disease and quantitative interaction, (b) rare disease and qualitative interaction, (c) common disease and quantitative interaction, and (d) common disease and qualitative interaction. Five testing methods are plotted: the case-control interaction (solid), the case-only interaction (short dashes), three two-stage procedures: marginal association screening and case-control interaction testing (long dashes), marginal association screening and case-only interaction testing (dot-dash), and gene-environment correlation screening and case-control interaction testing (dots).

(c)

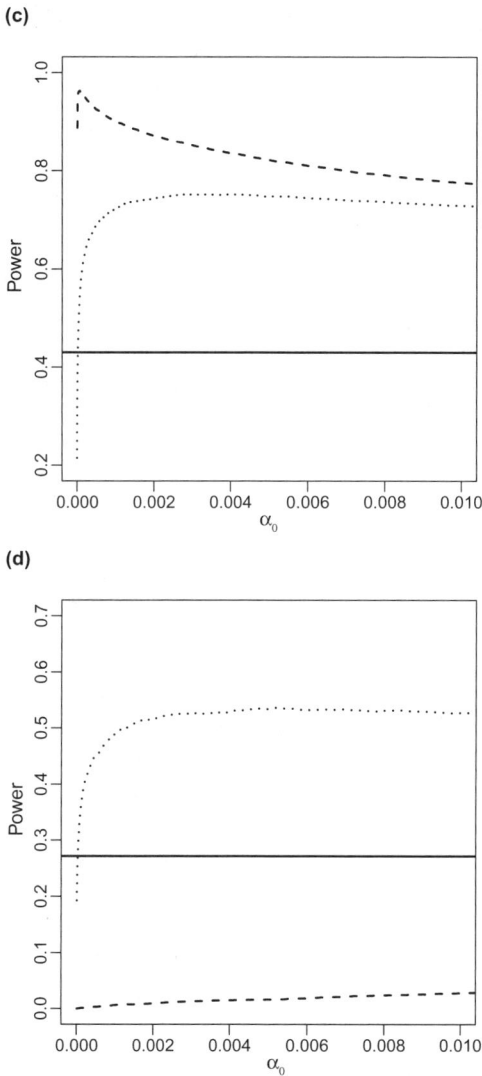

(d)

Figure 2.3 (continued)

is a quantitative interaction, filtering on the main genetic effect followed by a case-only analysis is the most powerful, and when the interaction is qualitative, a case-only analysis without screening is most powerful. When the disease is not rare (and a case-only analysis is thus no longer appropriate) a two-stage procedure filtering on marginal association followed by a case-control analysis is most powerful when the interaction is quantitative; filtering on gene-environment correlation followed by a standard case-control analysis is most powerful when the interaction is qualitative. We thus conclude that it depends on the type of interaction which procedure is most powerful; we also note that for the two-stage procedures there is some dependence on the level of stringency in screening during the first stage.

Combinations of Screening Statistics

This conclusion that it depends on the situation (e.g., type of interaction, research design) as to which procedure to identify interactions is most powerful, is rather unsatisfactory. A natural approach is to allocate the overall significance α for SNPs that pass the screening into two parts: $\rho\alpha$ for the gene-environment correlation screening G-E correlations and $(1 - \rho)\alpha$ for SNPs passing the genetic association screening with $0 \leq \rho \leq 1$ [9]. Although Murcray et al. set $\rho = 0.5$ in all their simulations, the choice of ρ is somewhat arbitrary. To improve the effectiveness of screening while gaining power to detect the interaction, two procedures that combine both screening test statistics have been proposed.

Cocktail Approach As described in the previous section, gene-environment correlation screening followed by case-only interaction testing would inflate type 1 error rate. In order to take advantage of the powerful case-only test, Hsu et al. [10] proposed combining two screening statistics:

$$p_i^{screen(I)} = p_i^{marg} Ind(p_i^{marg} \leq c) + p_i^{corr} Ind(p_i^{corr} > c),$$

and

$$p_i^{screen(II)} = min(p_i^{marg}, p_i^{corr}),$$

where $Ind(.)$ is an indicator function that is 1 if the argument is true and 0 otherwise, p_i^{marg} and p_i^{corr} are the p-values corresponding to the marginal association in model (2.3) and gene-environment correlation in model

(2.4) for the ith SNP, respectively, and c is a preselected constant. For each of these screening statistics the subsequent testing of G × E is done using the standard case-control estimator based on model (2.1) if $p_i^{screen} = p_i^{corr}$, while if $p_i^{screen} = p_i^{marg}$ the testing can be using the powerful case-only model (2.2) if there is G-E independence (assumed), or using the empirical Bayes estimator of Mukherjee and Chatterjee [11] that is a weighted average of the case-only and case-control estimators. Hsu et al. [10] established, using the framework of Dai et al. [2], that $p_i^{screen(I)}$ maintains the type 1 error, while $p_i^{screen(II)}$ is slightly conservative. Indeed, in most of their simulations $p_i^{screen(I)}$ appears slightly more powerful; however, the advantage of $p_i^{screen(II)}$ is that no constant c needs to be selected.

EDGxE Gauderman et al. [12] propose EDGxE, another combination of the two screening statistics from the marginal model (2.3) and the G-E correlation (2.4). In particular, let S_{dg} be the χ^2 statistics corresponding to the hypothesis $\gamma_1 = 0$ in (2.3), and let S_{ge} be the χ^2 statistics corresponding to the hypothesis $\delta_1 = 0$ in (2.4). To combine the strength of both screening statistics, the EDGxE method uses $S_{dg} + S_{ge}$ to select the SNP on which to test G × E. The G × E testing uses the case-control estimator in (2.1), as the EDGxE approach uses G-E correlation in the screening, thus use of the case-only estimator would not maintain the type 1 error. Gauderman et al. [12] cited Dai et al. [2] for deriving the independence of S_{dg} and S_{ge}, and so the use of $\chi^2(2)$ distribution for the sum of the two statistics. This is a misinterpretation of the results in Dai et al. [2], who stated the G-E correlation is independent of the adjusted genetic effect, expressed as β_{adj} in

$$\text{logit}(Y = 1 \mid G, E) = \beta_0 + \beta_{adj}G + \beta_2 E, \tag{2.5}$$

because the entire distribution of G and E, including the G-E correlation in (2.4), are conditioned in the regression model (2.5). The quantity S_{dg}, which is based on marginal association, is not independent of the correlation-based S_{ge}. Use of S_{dg}, rather than the proper χ^2 statistic derived from the adjusted association β_{adj} in (2.5), does not affect the G × E test validity, since both S_{dg} and S_{ge} are independent of the case-control interaction estimator in (2.1). However, it may affect the power, since the ranking of top SNPs of the EDGxE method has not accounted for the correlation between the two screening statistics.

Comparisons of These Procedures We compared the power of these two hybrid screening statistics: Cocktail and EDG×E (figure 2.4). As a comparison, we also include case-control and empirical Bayes G × E interaction tests without screening, a marginal association screen followed by empirical Bayes testing and a G-E correlation screen followed by case-control testing.

Briefly, we consider a sample of n = 10,000 cases and 10,000 controls and a total of 1,000,000 independent SNPs. Both G and E are binary with frequency 0.5 and 0.3, respectively. The disease prevalence is 0.044. The genome-wide type 1 error rate is controlled at 0.05. For the case-control and empirical Bayes G × E tests without screening, the type 1 error per SNP is controlled at 5×10^{-8} = 0.05/1,000,000. For the two-stage procedures, we calculate the expected p-value of each screening statistic analytically. More specifically, under model (2.1), we calculate the expected log-odds ratio and the standard error for the marginal association of G with disease risk and the association of E with G under the case-control sampling. We also derive the joint distribution of the Z statistics of marginal association and correlation screening, and from this we obtain the proper p-values for the cocktail and EDG×E screening. For a given threshold for screening (here, we selected the threshold corresponding to significance level 0.001 in the screening stage), we calculate the probability of each screening statistic exceeds the threshold. We then compute the power for G × E testing with per SNP type 1 error controlled at 0.05/expected SNPs that pass the threshold, and multiply it by the probability of the screening statistic exceeds the threshold, which is the final power for two-steps procedure. The empirical Bayes (EB) estimator is used when the screening statistic is based on marginal association and cocktail to account for potential G and E correlation. An R package for power calculations for two-stage testing (powerGWASinteraction) is available from CRAN (http://cran.us.r-project.org/).

The two hybrid screening methods, Cocktail and EDG×E, show consistently good power compared to all other approaches under all scenarios considered. Both the conventional case-control and the empirical Bayes (EB) G × E estimators can be very powerful under some scenarios, but can lose substantial power under others scenario. That appears to be the case for marginal screening and correlation screening as well.

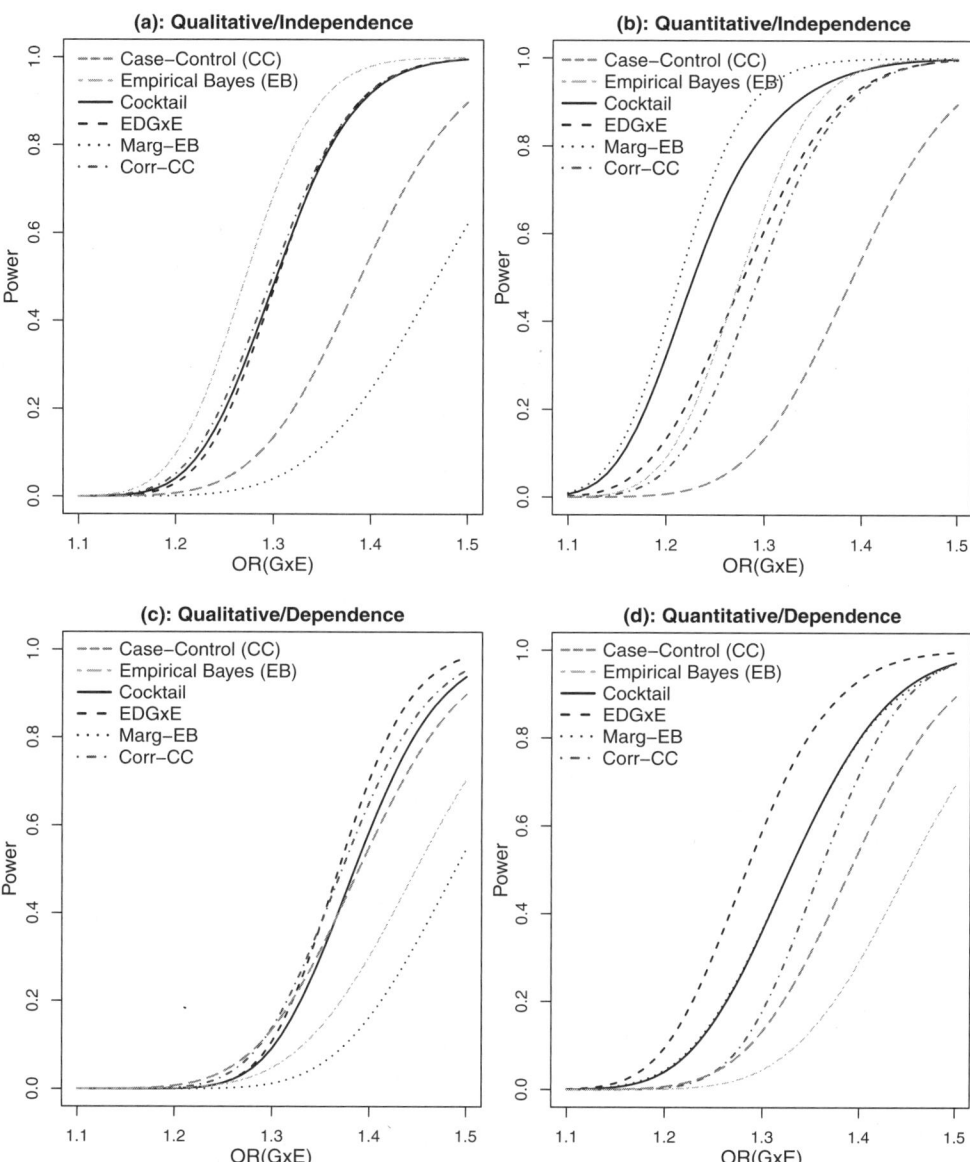

Figure 2.4

Power comparison of two hybrid screening statistics: Cocktail and EDG×E, as well as case-control (CC) and empirical Bayes (EB) G × E testing without screening and two screening statistics: marginal association (Marg-EB) and G-E correlation (Corr-CC). (a) and (b) are under qualitative and quantitative interaction, respectively and for both independent and dependent G and E. The OR of the main effect of E is 1.2. The OR of the main effect of G is 1.05 under the quantitative interaction model and 0.95 under the qualitative model. (c) and (d) are similar to (a) and (b) except that G and E are correlated with odds ratio 0.95. Note that in panel (d) the Marg-EB and Cocktail lines are virtually on top of each other.

Practical Issues

Use of Weighted Hypothesis Testing Framework to Account for the Screening Information

The essence of two-stage procedures is to select only a subset of SNPs for G × E testing and reduce the substantial burden of multiple comparison adjustment from testing of millions of SNPs for G × E to hundreds of SNPs. A common approach is to set a priori a threshold for the screening statistics. As shown in figure 2.3, the power of the two-stage procedures depends on the threshold value of the screening statistics.

We can formalize the two-stage testing by so-called weighted hypothesis procedures [13, 14]. Weighted procedures multiply the threshold by the weight w, for each test, raising the threshold when $w >1$ and lowering it if $w <1$. For example, testing for 1 million SNPs, the Bonferroni correction for each of the 1 million SNPs sets the threshold to be $\alpha/1,000,000$. Each test receives an equal weight, 1, and the average weight is 1. For two-stage procedures, if a hundred SNPs are selected for testing G × E, then for these 100 SNPs, the threshold is $\alpha/100$. Compared to the Bonferroni correction for 1,000,000 SNPs, the weight for these 100 SNPs is 10,000, but for the rest of the SNPs, the weight is 0, that is, the threshold is infinity, which yields an average weight to be 1.

If the weights are informative, the procedure improves power substantially; however, if the weights are uninformative, there is potential power loss. The screening statistics that are proven to be independent of G × E tests provide a natural basis for selecting the weights. A straightforward two-stage procedure is a weighted hypothesis test with a binary weight, either a constant for selected SNPs or 0 for nonselected SNPs. A more general form of the weight could have different weights depending on the value of the screening statistic. For example, we could select a piecewise constant weight function where, depending on the ranking of the SNPs among the screening statistics, a differential weight can be assigned to each SNP.

Ionita-Laza et al. [15] propose a strategy by partitioning SNPs into a relatively small number of partitions such that SNPs that belong to the same partition receive the same weight and the SNPs ranked high are given greater weight. According to their strategy, let K be the initial group size; then the first group consists of the top K ranked SNPs, the second group consists of the next $2 \times K$ ranked SNPs, the third group consists of the following $2^2 \times K$ ranked SNPs, and so on. The jth group would consist of 2^{J-1} $\times K$ SNPs that are ranked between $(2^{J-1}-1) \times K + 1$ and $(2^J-1) \times K$. The corresponding group-level significance thresholds are $\alpha/2$, $\alpha/4$, $\alpha/8$, ... and the overall type 1 error is then $\alpha/2 + \alpha/4 + \alpha/8 + ... < \alpha$. Within each group, the individual SNP significance threshold is calculated using a Bonferroni correction. For example, for the jth group, the group-level threshold is $\alpha/2^J$; since there are $2^{J-1} \times K$ in the jth group, the SNP-level threshold is $(\alpha/2^J)/ (2^{J-1}$ $\times K)$. Taking 1,000,000 SNPs for G × E testing and an initial group size $K = 5$, this results in the weight for the SNPs in the first 9 groups, a total of 2,555 SNPs, to be larger than 1, i.e., less stringent threshold for claiming statistical significance while the rest of the SNPs have weight < 1; a more stringent threshold.

We examine the number of SNPs being selected based on the expected p values of each screening statistic under model (2.1). This provides some insight into how the weight may be assigned to the SNPs. As an illustration, we focus on the scenario of a quantitative interaction with no correlation between G and E for two different sample sizes of $n = 20,000$ and 40,000, half cases and half controls (figure 2.5). The ORs of the main effects of G

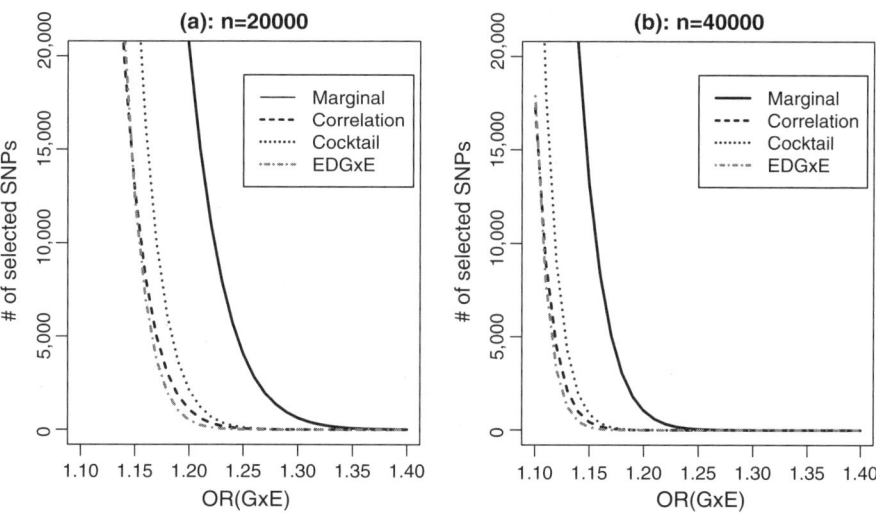

Figure 2.5
The number of SNPs being selected in order to include the SNP with G × E under the quantitative interaction model with no correlation between G and E in the population. Panels (a) and (b) are for 20,000 and 40,000 cases and controls, respectively.

and E are assumed to be 1 and 1.2, respectively. When there are $n = 20,000$ cases and controls, in order for the SNP with G $\times$ E to be among the top 2,500 SNPs with probability 0.5, the minimum OR for G $\times$ E interaction is 1.27 for the marginal association screening and 1.18~1.20 for the correlation, cocktail, and EDG×E screening. When $n = 40,000$, the minimum OR for G $\times$ E interaction is reduced to 1.19 for the marginal association screening and 1.13~1.14 for correlation, cocktail, and EDG×E screening. Interestingly, despite that a larger number of SNPs needs to be selected for the marginal association screening, coupling with the case-only G $\times$ E test the marginal association screening remains as powerful as the correlation and the EDG×E screening (results not shown). This can be partly explained that in the scenario of G and E independent in the population, the G $\times$ E interaction is approximately equivalent to the correlation of G and E in cases only. Such information, whether it would be used as a correlation-screening statistic or in a G $\times$ E testing with the marginal association screening, can improve power substantially.

While we show that the G $\times$ E interaction can induce correlation between G and E and marginal association of G with disease risk, there are SNPs that are correlated with E or significantly associated with disease risk without interaction. These SNPs would also be ranked high, and potentially lower the ranking of SNPs with G $\times$ E interaction, which is the main interest here.

In conclusion, if the sample size of a study is limited, it may be advisable to choose a large K in the weighted hypothesis testing instead of focusing on only a few top SNPs so that SNPs with G $\times$ E would not be missed. Depending on the scenarios and the G $\times$ E procedure, the value of K may vary.

Correlated SNPs (LD)

The dense genome-wide markers are correlated spatially, that is, SNPs that are close in proximity are likely to be correlated and share the same ancestral chromosomes. This nonrandom association between SNPs is called linkage disequilibrium (LD). Correlated SNPs will have similar strength of associations with outcome and other covariates. This poses a practical issue when ranking the SNPs, because if we ranked the SNPs simply by their screening statistics, the correlated SNPs would be clustered together and assigned similar weight, leaving little chance for other independent SNPs to be ranked on the top.

A simple and quick approach is to do distance pruning. If a SNP is within a prefixed distance, say 50KB, of another SNP that is ranked higher, the SNP is removed from the list. A more refined pruning is by the measure of LD; however, calculating pairwise LD for millions of SNPs remains a daunting task.

Randomized Clinical Trial (e.g., Treatment)

Sometimes the environmental effect of interest is studied in randomized clinical trials. Such trials provide an ideal setting for studying gene-environment interactions, specifically gene-treatment interaction for a number of reasons. First, the treatment regime can be controlled and prescribed differentially according to individual genetic background; therefore, discovering gene-treatment interaction can exert an immediate public health impact. Second, as compared to observational studies, the unequivocal gene-treatment independence by design facilitates more efficient and robust estimation procedures. Extensions of the case-only estimator have been proposed for randomized clinical trials with a rare endpoint and nondifferential censoring across arms [16, 17]. Notably, both subgroup treatment effects and interactions can be estimated from (2.2) when an offset—the logarithm of the randomization ratio—is added to the linear predictor. The other less appreciated benefit from randomization is the robustness of gene-treatment interaction to unmeasured confounding [18]. Unmeasured confounding variables can lead to biased estimates of gene-environment interaction in observational studies. In randomized trials, however, there should be no confounding for treatment effect, and confounding of genetic effect is often largely eliminated by adjusting for population substructure, e.g., top principal components from the covariance matrix of genetic data from genome-wide association studies. Even if there is residual confounding of genetic effect, so that the interaction estimator is biased, testing for gene-treatment interaction remains valid under the null hypothesis [19].

The sample size of a randomized clinical trial is typically dictated by the primary hypothesis on treatment effect. Ancillary case-control genetic studies for gene-treatment effect are almost always severely underpowered, with less chance of finding appropriate meta-analysis since the number of treatment trials studying the same treatment is usually limited. Hence, there is a greater need to incorporate the two-stage testing procedures in discovering gene-treatment interaction. For example, in the hormone therapy trials of the Women's Health Initiative, we conducted a genetic association study to discover the gene-treatment interaction among patients with cardiovascular disease ($n = 517$), stroke ($n = 349$), VTE ($n = 313$) and type-2 diabetes ($n = 1,078$), as well as 1:1 matched controls. This sample size is clearly insufficient for any small to moderate interaction effect size. We have adopted the aforementioned two-stage procedures with modifications. As shown in Dai et al. [2], the estimate of the marginal genetic effect in a logistic regression model is orthogonal to the case-only estimator of gene-treatment interaction, when trial endpoints are rare. We used the marginal effect screening and tested for interaction by case-only estimators. For common clinical endpoints, Dai et al. [17] used the semiparametric likelihood estimator under two-phase sampling to exploit gene-treatment independence, thereby generalizing the case-only estimator. Initial simulation results suggest that the efficient gene-treatment interaction estimator derived by Dai et al. [2] may also be orthogonal to the usual marginal genetic effect, leading to the possibility of using such efficient estimator in the second stage testing. The advantage is that the rare disease assumption is no longer required for the testing stage to use efficient estimators.

The standard analytical framework for randomized trials with a failure time endpoint is the Cox proportional hazards model, and so the treatment effect is typically expressed in hazards ratio. Dai et al. [2] showed that the case-only estimator obtained from the usual logistic regression model (2.2) can be interpreted as hazard ratio for subgroup treatment effect. Furthermore, the marginal genetic effect in a Cox model was shown to be orthogonal to the case-only gene-treatment interaction. These results extend the two-stage testing framework to survival analysis where genotypes are potentially ascertained in a case-cohort sample.

Form of Interaction (Additive vs Multiplicative)

We have so far focused on the multiplicative interaction because the logistic regression model has been the workhorse of contemporary epidemiology. Alternatively, additive interaction has been advocated by some epidemiologists, in that the latter is believed to be more relevant for assessing the biological interaction and more interpretable in evaluating public health impact [20]. The estimation of additive interaction under case-control sampling, however, is not yet well developed, especially for disease outcomes that are not rare, since retrospective sampling cannot be conveniently ignored in model fitting, as is the case of testing multiplicative interaction. The other roadblock is covariate adjustment in the additive scale, as continuous covariates can easily impose model misspecification. Recently there have been several interesting developments to address these problems [19, 21]. Our discussion herein pertains to whether the additive interaction can be used in the second-stage testing after the first-stage screening or weighting.

Suppose we used case-control sampling to genotype a subset of participants in a cohort study or a randomized trial, and hence we know the sampling fraction. The additive model for assessing gene-environment interaction is

$$E(Y = 1 \mid G, E) = \alpha_0 + \alpha_1 G + \alpha_2 E + \alpha_3 GE,$$

assuming G and E are both binary. If G and E are observed in the entire cohort, theorem 2 of Dai et al. [2] applies immediately and the estimator α_3 is orthogonal to the marginal genetic effect in a logistic model. Under the case-control sampling, we can weight the estimating equation for the additive interaction model by the inverse of case-control sampling fractions to attain unbiased estimation. A straightforward extension of Dai et al. [2] suggests that the weighted estimator for the additive interaction will be orthogonal to the marginal effect in the logistic model as well. Since the G-E correlation does not use disease information, the independence of G-E correlation and the additive interaction is immediate. The various two-stage approaches we discussed previously will be applicable to additive interactions.

The inverse probability weighted estimators are typically inefficient but robust. In the simple scenario we consider here, e.g., binary G and binary E, it is useful to pursue more efficient estimator for additive interaction, for example the maximum likelihood estimator under two-phase sampling in the same vein as Dai et al. [17]. Further development is needed to extend Han et al. [21] to exploit gene-treatment independence for additive interaction in randomized trials, and to incorporate covariate adjustment into additive interaction models in a robust manner.

Gene-Gene Interactions

The two-stage screening testing procedures that we described for gene-environment interactions clearly can also be used for gene-gene (G × G) interactions. In fact, the proposal made in Kooperberg and LeBlanc [3] was made in the context of G × G interactions. Here we discuss a few issues that are unique to G × G interactions.

Because of LD, many (nearby) SNPs are highly correlated. As a result, a correlation based screen would make very many pairs of SNPs pass the first stage just because they are in LD, rather than because they are good candidates for having a G × G effect on the phenotype. In fact, Lewinger et al. [8], who proposed a two-stage procedure for identifying G × G interactions, suggested an adjustment for ancestry to reduce the false positives. This would presumably address these issues, if the only source of LD is ancestral—which may be the case for spurious correlations between SNPs on different chromosome, but not necessarily fully control for correlations between SNPs in the same haplotype.

For G × E interactions, we typically screen only on the SNPs, and the screening is unambiguous. For example, the marginal association screening would be the Z-statistic in model (2.3), which we here will refer to as $Z(G)$. For an interaction between two SNPs, say G_1 and G_2, there are many different ways that we can combine the two Z-statistics and still get a valid screening procedure. The two most obvious screening statistics are $\min(|Z(G_1)|, |Z(G_2)|)$, which was used in Kooperberg and LeBlanc [3] and $\max(|Z(G_1)|, |Z(G_2)|)$, that is, only one SNP has to pass the screening stage for the pair to be tested. However, other combinations could be considered, e.g., $Z(G_1)^2 + Z(G_2)^2$, so that stronger evidence for one SNP could compensate for lack of evidence for the other SNP. Each of those rules would still satisfy the conditions of Dai et al. [2].

Continuous Outcomes and Environmental Factors

We have described the methods to identify G × E and G × G interactions in the context of case-control studies. We note, however, that some of the most successful GWAS have been for continuous phenotypes, such as height and lipid levels (see the NHGRI GWAS catalog: www.genome.gov/gwastudies). The methodology that we proposed for two-stage testing for interactions carries over to such outcomes; in particular, models (2.1) and (2.3) can simply be replaced by the usual linear regression models. The proposed methods are applicable for continuous environmental factors; in this situation, model (2.4) is also replaced by a linear regression model.

Concluding Remarks

Genome-wide association studies and next generation sequencing studies offer us an unprecedented opportunity to study the genetic etiology of diseases and other traits. Over the last few years, many replicated associations between SNPs and traits have been published. It is of particular interest to identify how genes may interact with environmental factors and other genes. In this chapter, we show that a two-stage approach, where in the first stage SNPs are screened for their potential to be involved in interactions, and interactions are then tested only among SNPs that pass the screening can greatly enhance power for detecting gene-environment and gene-gene interaction in large genetic studies compared to the tests without screening. There are many open questions, such as the optimal choice of weights when using weighted hypothesis testing, noting that these weights could be based on the screening statistics, accounting for correlated SNPs, and heterogeneity among studies in the meta-analysis. Future research along these lines may further improve the effectiveness and broad applicability of the two-stage approaches to the study of G × E and G × G interactions.

Acknowledgments

This work was supported in part by National Institute of Health grants P01 CA053996, R01 AG014358, R01 HG006124, R01 HL114901, U01 CA137088, and R01 CA059045.

References

1 Smith, P. G., & Day, N. E. (1984). The design of case-control studies: The influence of confounding and interaction effects. *International Journal of Epidemiology, 13,* 356–365.

2. Dai, J. Y., Kooperberg, C., LeBlanc, M., & Prentice, R. L. (2012). Two-stage testing procedures with independent filtering for genome-wide gene-environment interaction. *Biometrika, 99,* 929–944.

3. Kooperberg, C., & LeBlanc, M. (2008). Increasing the power of identifying gene × gene interactions in genome-wide association studies. *Genetic Epidemiology, 32,* 255–263.

4. Millstein, J. (2013). Screening-testing approaches for gene-gene and gene-environment interactions using independent statistics. *Frontiers in Genetics, 4,* 306.

5. Kooperberg, C., LeBlanc, M., Dai, J. Y., & Rajapakse, I. (2009). Structures and assumptions: Strategies to harness gene × gene and gene × environment interactions in GWAS. *Statistical Science, 4,* 472–488.

6. Murcray, C. E., Lewinger, J. P., & Gauderman, W. J. (2009). Gene-environment interaction in genome-wide association studies. *American Journal of Epidemiology, 169,* 219–226.

7. Millstein, J., Conti, D. V., Gilliland, F. D., & Gauderman, W. J. (2006). A testing framework for identifying susceptibility genes in the presence of epistasis. *American Journal of Human Genetics, 78,* 15–27.

8. Lewinger, J. P., Morrison, J. L., Thomas, D. C., Murcray, C. E., Conti, D. V., Li, D., et al. (2013). Efficient two-step testing of gene-gene interactions in genome-wide association studies. *Genetic Epidemiology, 37,* 440–451.

9. Murcray, C. E., Lewinger, J. P., Conti, D. V., Thomas, D. C., & Gauderman, W. J. (2011). Sample size requirements to detect gene-environment interactions in genome-wide association studies. *Genetic Epidemiology, 35,* 201–210.

10. Hsu, L., Jiao, S., Dai, J. Y., Hutter, C., Peters, U., & Kooperberg, C. (2012). Powerful cocktail methods for detecting genome-wide gene-environment interaction. *Genetic Epidemiology, 36,* 183–194.

11. Mukherjee, B., & Chatterjee, N. (2008). Exploiting gene-environment independence for the analysis of case-control studies: An empirical Bayes-type shrinkage estimator to trade-off between bias and efficiency. *Biometrics, 64,* 685–694.

12. Gauderman, W. J., Zhang, P., Morrison, J. L., & Lewinger, J. P. (2013). Finding novel genes by testing G × E interactions in a genome-wide association study. *Genetic Epidemiology, 37,* 603–613.

13. Genovese, C. R., Roeder, K., & Wasserman, L. (2006). False-discovery control with *p*-value weighting. *Biometrika, 93,* 509–524.

14. Roeder, L., & Wasserman, L. (2009). Genome-wide significance levels and weighted hypothesis testing. *Statistical Science, 24,* 398–413.

15. Ionita-Lazza, I., McQueen, M. B., Laird, N. M., & Lange, C. (2007). Genomewide weighted hypothesis testing in family-based association studies, with an application to a 100K scan. *American Journal of Human Genetics, 81,* 607–614.

16. Vinttinghoff, E., & Bauer, D. C. (2006). Case-only analysis of treatment-covariate interactions in clinical trials. *Biometrics, 62,* 769–776.

17. Dai, J. Y., Kooperberg, C., & LeBlanc, M. (2009). Semiparametric estimation exploiting covariate independence in two-phase randomized trials. *Biometrics, 65,* 178–187.

18. VanderWeele, T. J., Mukherjee, B., & Chen, J. (2012). Sensitivity analysis for interactions under unmeasured confounding. *Statistics in Medicine, 31*, 2552–2564.

19. VanderWeele, T. J., & Vansteelandt, S. (2011). A weighting approach to causal effects and additive interaction in case-control studies: Marginal structural linear odds model. *American Journal of Epidemiology, 174*, 1197–1203.

20. Rothman, K. J., Greenland, S., & Walker, A. M. (1980). Concepts of interaction. *American Journal of Epidemiology, 112*, 467–470.

21. Han, S. S., Rosenberg, P. S., Garcia-Closas, M., Figueroa, J. D., Silverman, D., Chanock, S. J., et al. (2012). Likelihood ratio test for detecting gene-environment interactions under an additive risk model exploiting G-E independence for case-control data. *American Journal of Epidemiology, 176*, 1060–1067.

3 Marker-Set Approaches for Assessing Gene × Environment Interactions at Gene Level

Jung-Ying Tzeng and Arnab Maity

Human complex traits have a multifactor etiology that involves interplay between genetic susceptibility and environmental exposures. Studies on gene-environment interactions (G × E) can facilitate the understanding of genetic heterogeneity under different environmental exposures [1, 2], help to identify high-risk subgroups in the population [3], provide insights into the biological mechanisms of complex diseases [4], uncover hidden heritability [5, 6], and improve the ability to discover susceptible genes that interact with other factors but exhibit small marginal effects [4].

Conventionally, researchers have searched for significant genetic or G × E associations using single-SNP methods. Many statistical methods for evaluating SNP × E effects have been developed (see, e.g., [7] for a comprehensive review). Recently, there is increasing interest in studying G × E effect at gene level. Several factors motivate the paradigm shift from SNP level to gene level. First, genes are the basic units in the biological mechanism and SNPs within a gene tend to work concordantly. Thus, assessing the modification effect of a gene under different environmental exposures may yield results that are more biologically insightful, easier to interpret, and more informative in revealing underlying mechanisms than SNP-level results. Second, gene-level analysis can naturally account for interactive effects among SNPs and overcome etiological heterogeneity. Finally, a gene-level analysis incorporates linkage disequilibrium (LD) information from all SNPs simultaneously within the gene. Consequently, such joint analysis of SNPs should have improved ability to tag untyped causal variants compared to the analysis of individual SNPs. Finally, the polygenic nature of complex diseases suggests moderate effect sizes for individual variants. Joint analysis of SNPs in aggregate can be more powerful than separate analysis of each individual SNP. The potential power gain can be from the amplification of

genetic signals, the reduction of degrees of freedom, and even the allevia-
tion of multiple-testing penalty when a whole-genome search is performed.

Gene-Based G × E Approaches for Common Variants

Several gene-based G × E methods are available for common variants, where
the major task is to avoid a large number of parameters for modeling genetic
variables (G), environmental variables (E), and G × E variables. Depending
on how the G × E effect is modeled, we roughly classify these approaches into
fixed-effects methods and random-effects methods. We review the major
methods of each category in this section. However, because the fixed-effects
approaches are sensitive to model misspecification, as discussed at the end
of the following subsection, we introduce the random-effects approaches in
further detail than the fixed-effects approaches.

We define the notation used in this chapter. Let Y_i be the trait values,
which can be continuous or binary. Let G_{mi} be the minor allele count for
individual i at locus m ($m = 1,...,M$), let X_{Ei} be a $1 \times K_E$ vector of environmental factors, and let X_{Ci} be the $1 \times K_C$ vector of confounders. The full
covariate vector is $X_i = (1, X_{Ci}, X_{Ei})$ with dimension $1 \times (1 + K_C + K_E)$.

Fixed-Effects G × E Approaches

Roughly speaking, the fixed-effect methods reduce the G × E degrees of
freedom (df) by obtaining a certain weighted sum of the genotypes (often
referred to as mutation burden) that summarizes the multi-marker information. Then, based on this burden genotype term, a G × E design vector
can be defined and used for assessing the G × E effect at the gene level. The
weights can be obtained from the signal size of main effects [8], from the
use of principal components (PCs) or partial least square (PLS) models of
the SNP genotypes [9], or from the strength of G-E correlations [10].

Tukey's one degree-of-freedom test [8] is one of the first proposed G × E
marker-set methods. Motivated by a latent-variable model where the
observed SNPs are considered as a surrogate of the (latent) underlying biological phenotypes affecting the disease risk, Chatterjee and colleagues
showed that the interactions among the SNPs of a gene and a group of
environmental factors (e.g., $K_E > 1$) can be detected by the parsimonious
Tukey's one df test. Specifically, Tukey's 1-df interaction model assumes that
the interaction effect is proportional to the product of the main effect of

the E variable and the main effect of the gene, which are described by a weighted SNP sum with the weights being the SNP's effect on the trait:

$$g(\mu_i) = \beta_0 + X_{Ci}\beta_C + \sum_{\ell=1}^{K_E} \beta_{E\ell} X_{E\ell i} + \sum_{m=1}^{M} \beta_{Gm} G_{mi} + \theta \sum_{m=1}^{M} \sum_{\ell=1}^{K_E} \beta_{E\ell} \beta_{Gm} X_{E\ell i} G_{mi},$$

where μ_i is the conditional trait mean, $g(\cdot)$ is the link function, β_0 is the intercept, and β_v's are the regression coefficients for the effect of variable v. With $K_E = 1$, the model reduces to

$$g(\mu_i) = \beta_0 + X_{Ci}\beta_C + \beta_E X_{Ei} + \sum_{m=1}^{M} \beta_{Gm} G_{mi} + \theta \left(\underbrace{\sum_{m=1}^{M} \beta_{Gm} G_{mi}}_{\textit{weighted burden}} \right) \beta_E X_{Ei}.$$

The G × E effect can be evaluated by testing for $H_0 : \theta = 0$. This model contributed in making progress toward being able to understand complex diseases more fully using a few degrees of freedom; however, it relied on the often incorrect assumption that a SNP's interaction effect is proportional to its marginal genetic effect [11].

Principal component (PC)/partial least square (PLS) regressions [9] used certain dimension reduction tools such as PC or PLS to summarize the information of multiple G variables (or the multiple E variables), and then created the G × E terms using the cross product of the top components of the G variables (or the E variables). Unlike PCs that are computed solely from genotype data (and hence account for the LDs among SNPs), PLS aims to extract components that summarize both the LDs among SNPs and the correlation between SNPs and the traits. Wang et al. [9] showed that the PC/ PLS regressions often had better performance than the Tukey 1-df method, particularly when causal SNPs had no or negligible marginal effects. However, PC regressions have been found to have inflated type 1 error rates for G × E tests when there exist interactive effects among the SNPs or among the E variables [12]. Under the content of using PC regressions of G × G tests, Wang et al. [13] found that the cross-product of the PCs from each gene (e.g., $PC_{geneA} \times PC_{geneB}$) would capture less information than the PCs obtained from the G × E design matrix (i.e., $G_{mi} \times X_{Ei}$), and consequently lead to a higher false positive rates.

The set-based gene environment interaction test (SBERIA) method [10] aims to address two major challenges encountered in set-based tests: to distinguish signal SNPs from noise in a marker set and to determine the effect direction of the signals. Jiao et al. showed that the correlation between

G and E variables in the pooled case-control sample can provide information about the G × E effect and that such information is independent with the interaction test. Therefore, SBERIA incorporates G-E correlation as weights to increase the signal-to-noise ratio in a G × E set test while avoiding permutations. The SBERIA model, which assumes $K_E = 1$, has the form of $g(\mu_i) = \beta_0 + X_{Ci}\beta_C + \beta_E X_{Ei} + \sum_{m=1}^{M} \beta_{Gm}G_{mi} + \rho X_{Ei} \left(\sum_{m=1}^{M} w_m G_{mi} \right)$, where $w_m = I(|r_m| > \theta_n)sign(r_m)$, r_m is the correlation between X_{Ei} and G_{mi}, and $\theta_n = o\left(n^{\frac{1}{2}} \right)$. The G × E test evaluates $H_0 : \rho = 0$. However, this is valid only when the true G-E correlation is in the same direction as the G × E interaction.

Major Concerns of the Fixed-Effects Methods For the fixed-effects approaches, the G × E tests would have inflated type 1 error rate when the model does not correctly reflect the true underlying G effects and E effects [14, 12]. This is because when the underlying nonadditive effects among SNPs (i.e., SNP × SNP effects) or among the E variables (i.e., E × E effects) are not adequately modeled, the G × E term in the fitted model absorbs these SNP × SNP and E × E effects and contributes to false G × E findings when there were no G × E interactions. Similar results have been observed by Voorman et al. [15], in which it was shown that severe inflation of type 1 error rate can occur in G × E-GWAS if the fitted model used for G × E screening does not correctly reflect the true underlying G effects and E effects. To address the issue, Voorman et al. [15] suggested a model-robust estimate of the variance, and Lin et al. [14] and Wang et al. [12] suggested a random-effect model for capturing the genetic main effect.

Random-Effects G × E Approaches
Several random-effects methods are available to model G × E interactions, including generalized linear mixed model (GLMM), similarity regression (SimReg), and kernel machine regression (KMR). Although motivated differently and constructed under different frameworks, all of these methods can be unified under GLMM. We will introduce each of the methods separately and illustrate the connection to GLMM for each method.

Generalized Linear Mixed Model (GLMM) The testing framework for interaction between an environmental variable and a set of genetic markers using generalized linear mixed model was proposed in Lin et al. [14]. They first demonstrated that the common practice of single-marker G × E analysis may be biased if the main effects of multiple SNPs are associated with the outcome and proposed a more sophisticated approach named gene-environment set association test (GESAT).

Lin et al. [14] considered having a scalar environment factor and simple one-way interactions between the markers and the environmental variables. Specifically, consider the mean model

$$g(\mu_i) = \beta_0 + X_{Ci}\beta_C + \beta_E X_{Ei} + \sum_{m=1}^{M} \beta_{Gm} G_{mi} + \sum_{m=1}^{M} \gamma_m X_{Ei} G_{mi},$$

where γ_m are the unknown interaction effects. It is also assumed that the response variables, conditional on the covariates, are generated from an exponential family.

Typical G × E models often analyze one SNP at a time, and thus they fit a misspecified model involving only one marker. In such a situation, Lin et al. [14] showed that if the true model is indeed a multimarker model, then under the null hypothesis of no interaction between the marker set and the environmental variable, that is, $H_0 : \gamma_m = 0, m = 1, \ldots, M$, the single marker G × E test is typically biased and produced incorrect type 1 error rates. This is due to the fact that the estimated interaction coefficients from the misspecified single marker models are generally biased. As a result, the standard approach of taking the minimum of the p-values of the single markers tests is generally invalid. It should be also noted that even if the interaction coefficients are unbiased (as may be the case when the marker set variables are independent of the environmental variable), standard inference can still be invalid because the estimated standard error estimates can be biased.

To overcome this issue, Lin et al. [14] developed the GESAT to test $H_0 : \gamma_m = 0, m = 1, \ldots, M$. The main idea behind this test is to consider that each γ_m follows an arbitrary distribution with zero mean and common variance τ^2, and that they are independent of each other. Thus, the null hypothesis of no interaction becomes the equivalent of testing the null hypothesis $H_0 : \tau^2 = 0$. Lin [16] developed a score testing procedure for such variance components testing. Following a similar technique, the test statistic is given by

$$T = (Y - \hat{\mu})^T S_{Int} S_{Int}^T (Y - \hat{\mu}),$$

where $Y = (Y_1, \ldots, Y_n)^T$, $\hat{\mu}$ is the estimate of μ under H_0, and S_{Int} is an $n \times M$ matrix of interactions where the i th row is given by $(X_{Ei}G_{1i}, \ldots, X_{Ei}G_{Mi})$. When the number of markers in the marker set is small, we can use standard maximum likelihood estimation to estimate the parameters β_{Gm}. However, when the number of markers M is large, we propose using ridge regression to estimate the parameters by incorporating an additional penalty term $\lambda \sum_{m=1}^{M} \beta_{Gm}^2$ with the negative log-likelihood function, where λ is a tuning parameter. This tuning parameter can be estimated using the generalized cross-validation (GCV) method [17]. Lin et al. [14] showed that under H_0, the test statistic follows a mixture of chi-squares distribution with different mixing coefficients that depend on the estimated tuning parameter. Thus, the p-value of the test can be calculated using the characteristic function inversion technique described in Davies [18].

There are a few main advantages of using such a variance components test for G × E testing. This approach allows one to borrow information across γ_m's. This test has been shown to be the locally most powerful test [16] under some regularity conditions. Also, this testing procedure is computationally efficient, since one needs only to fit the null model (with no interactions) to compute the test statistic.

Gene-Trait Similarity Regression (SimReg) for G × E Effects

The SimReg G × E framework was established by Tzeng et al. [19] and was extended to binary traits [20]. SimReg is inspired by the Haseman-Elston regression model from linkage analysis [21, 22] and haplotype similarity tests for regional association [23, 24]. In SimReg, one first quantifies the trait similarity T_{ij} and genetic similarity S_{ij} of the targeted gene for each pair of individuals i and j. One then regresses the trait similarity on the genetic similarity and detects gene-trait association by testing for the significance of relevant regression coefficients. The SimReg framework has been developed for quantitative traits, binary traits [25], and survival traits [26].

The SimReg Model The trait similarity T_{ij} is based on the covariance between Y_i and Y_j by taking the product of the trait residuals of subjects i and j with some weights, e.g., $T_{ij} = \{\omega_i(Y_i - \mu_i^0)\}\{\omega_j(Y_j - \mu_j^0)\}$, where $\mu_i^0 = E(Y_i|X_i) = g^{-1}(X_i\gamma)$ is the subject-specific trait mean accounting for covariates X_i with effect γ but assuming no genetic effects, and ω_i is a weight accounting for the fact that the Y_i's have difference variances [25]. With this definition, the expected trait similarity value $E(T_{ij}X) = \omega_i\omega_j \times E\{(Y_i - \mu_i^0)(Y_j - \mu_j^0)\}$ is the covariance of Y_i and Y_j with weights $\omega_i\omega_j$. For continuous traits with a linear model, $\mu_i^0 = g^{-1}(X_i\gamma) = X_i\gamma$ and $\omega_i = 1$; for binary traits with a logistic model, $\mu_i^0 = g^{-1}(X_i\gamma) = e^{X_i\gamma}/(1 + e^{X_i\gamma})$ and $\omega_i = \mu_i^0(1 - \mu_i^0)$, which is the optimal weight for the logistic model [25].

The genetic similarity S_{ij} is measured by the sum of the weighted single-marker similarity score, i.e., $S_{ij} = \sum_{m=1}^{M} w_m s(G_{mi}, G_{mj})$, where $s(G_{mi}, G_{mj})$ is the genetic similarity at marker m and w_m is the weight. There are several choices for $s(G_{mi}, G_{mj})$ [27, 28]; a popular one is the identity-by-state (IBS) metric with allele frequency weights that can be calculated as follows. Define that g_i^m the allele count vector of marker m and subject i, where the length of g_i^m is the number of distinct alleles at marker m. For example, for a triallelic locus, $g_i^m = (1,0,1)^T$ if subject i has genotype A_1A_3 and $g_i^m = (0,2,0)^T$ if subject i has genotype A_2A_2. Then the IBS between subjects i and j and marker m is $s_{IBS}(G_{mi}, G_{mj}) = 2$ if $|g_i^m - g_j^m|$ is a zero vector; $s_{IBS}(G_{mi}, G_{mj}) = 1$ if $|g_i^m - g_j^m|$ contains exactly two 1's; and $s_{IBS}(G_{mi}, G_{mj}) = 0$ otherwise. Weights w_m are specified to upweight or downweight a variant based on certain features. Examples include weights that based on allele frequencies, the degree of evolutionary conservation, or the functionality of the variations [27, 28, 29, 30]. For example, one can use the minor allele frequency of marker m, denoted by q_m, to up-weight similarities that are contributed by rare variants: e.g., $w_m = (1 - q_m)^{24}$ [31] can be used to target rare variants only, or a moderate weight $w_m = q_m^{-3/4}$ [32] can be used to promote similarity attributed to rare alleles while retaining the contribution from common variants.

The proposed G × E gene-trait similarity regression model is as follows:

$$E(T_{ij}|X,S) = a + b \times X_{Ei}X_{Ej} + c \times S_{ij} + d \times S_{ij} \times X_{Ei}X_{Ej}, \quad i \neq j. \tag{3.1}$$

Because T_{ij} has incorporated baseline covariate information, model (3.1) does not contain an intercept or $X_{Ei}X_{Ej}$ covariate term (i.e., $a = b = 0$) [19]. With model (3.1), one can assess G × E interaction by testing $H_0^{GE} : d = 0$, or perform a joint test for the genetic main effect and G × E interactions simultaneously by testing $H_0^{Joint} : c = d = 0$. The joint test is recommended if either the genetic heterogeneity or G × E interaction mechanism is unknown [1, 19].

Connection of SimReg with GLMM It can be shown that the similarity regression can be represented as a working generalized linear mixed model (GLMM) [25, 20] as follows:

$$g(\mu) = X\gamma + h_G + h_{GE}, \tag{3.2}$$

where $\mu = (\mu_1, \ldots, \mu_n)$ is the vector of the conditional means with $\mu_i = E(Y_i | X, h_G, h_{GE})$, and $g(.)$ is the link function. Vectors $h_{G(n \times 1)} = (h_{G1}, \cdots, h_{Gn})$ and $h_{GE(n \times 1)} = (h_{GE1}, \cdots, h_{GEn})$ are the subject-specific genetic main effect and $G \times E$ interaction, respectively. In this working model, if we assume that h_G and h_{GE} are random effects, i.e., $h_G \sim N(0, \tau_G S_G)$ and $h_{GE} \sim N(0, \tau_{GE} S_{GE})$ with $S_G = \{S_{ij}\}$, $S_{GE} = DS_G D$, and $D = diag\{X_{Ei}\}$, then the marginal covariance of Y_i and Y_j is

$$cov(Y_i, Y_j) \approx \{g'(\mu_i^0) g'(\mu_j^0)\}^{-1} \times \{\tau_G S_{ij} + \tau_{GE} X_{Ei} X_{Ej} S_{ij}\},$$

where $g'(\mu) = \partial g(\mu) / \partial \mu$. Recall that the expected trait similarity is $E(T_{ij} X) = \omega_i \omega_j \times cov(Y_i, Y_j)$. Therefore,

$$E(T_{ij} X) \approx \omega_i \omega_j \times \{g'(\mu_i^0) g'(\mu_j^0)\}^{-1} \times \{c \times S_{ij} + d \times X_{Ei} X_{Ej} S_{ij}\}$$

$$= c \times S_{ij} + d \times X_{Ei} X_{Ej} S_{ij},$$

where $\omega_i = g'(\mu_i^0)$ is $\dfrac{1}{\mu_i^0(1 - \mu_i^0)}$ with logit link and is 1 with identity link. In other words, we can examine $H_0^{GE} : d = 0$ and $H_0^{Joint} : c = d = 0$ of model (3.1) by testing $H_0^{GE} : \tau_{GE} = 0$ and $H_0^{Joint} : \tau_G = \tau_{GE} = 0$ in model (3.2) respectively.

Score Tests for G × E Effects The connection of SimReg and GLMM greatly facilitate the derivation of the score test statistics of SimReg. That is, following Zhang and Lin [33], Tzeng et al. [19], and Zhao et al. [20] showed that the score statistic to examine the $G \times E$ effect (i.e., testing $H_0^{GE} : \tau_{GE} = 0$) can be obtained as

$$T_{GE} = \frac{1}{2} \left\{ (y_1^W - X^3)^T V_1^{-1} S_{GE} V_1^{-1} (y_1^W - X^3) \right\} \Big|_{\tau_G = \hat{\tau}_G, \tau_{GE} = 0, \gamma = \hat{\gamma}}$$

where $y_1^W = X\hat{\gamma} + Z_G \hat{b} + \Delta_G (y - \hat{\mu}^G)$ is the working vector in model (3.2) under $H_0^{GE} : \tau_{GE} = 0$; $\mu^G = E(Y | X, b) = g^{-1}(X\gamma + Z_G b)$; $\Delta_G = diag\{g'(\mu_i^G)\}$, and μ_i^G is the ith entry of μ^G; $\hat{\tau}_G$ and $\hat{\gamma}$ are the maximum likelihood estimates (MLE) for τ_G and γ under H_0^{GE} respectively; $V_1 = W_G^{-1} + \tau_G S_G$ with $W_G = diag\{\mu_i^G(1 - \mu_i^G)\}$, and $P_1 = V_1^{-1} - V_1^{-1} X (X^T V_1^{-1} X)^{-1} X^T V_1^{-1}$. In a similar manner, the score test statistic under $H_0^{Joint} : \tau_G = \tau_{GE} = 0$ is

$$T_{Joint} = \frac{1}{2}\{(y_0^W - X\gamma)^T V_0^{-1}(S_{GE} + S_G)V_0^{-1}(y_0^W - X\gamma|_{\tau_G=0, \tau_{GE}=0, \ \gamma=\tilde{\gamma}},$$

where $y_0^W = X\tilde{\gamma} + \Delta(y - \hat{\mu}_0)$ is the working vector defined as in T_{GE} except under $H_0^{joint}: \tau_G = \tau_{GE} = 0$, $\mu^0 = E(Y \mid X) = g^{-1}(X\gamma)$, $V_0 = W_0^{-1}$, $W_0 = diag\{\mu_i^0 (1 - \mu_i^0)\}$, $P_0 = V_0^{-1} - V_0^{-1}X(X^TV_0^{-1}X)^{-1}X^TV_0^{-1}$ and $\tilde{\gamma}$ is the MLE for γ under H_0^{Joint}. It can also be shown that T_{GE} and T_{Joint} follow a weighted χ^2 distribution asymptotically under H_0^{GE} and H_0^{Joint}, respectively [34, 19, 20]. The p-values can then be calculated numerically using moment-matching approximations [35] such as Davies [18].

Kernel Machine G × E Regression

Kernel machine regression (KMR) has emerged as an efficient and useful tool in association studies where one intends to investigate the joint association of a marker (or gene) set on an outcome of interest, especially when the number of genetic variables is large and there might be complex nonlinear effects. Kernel machine techniques were originally developed in the machine learning literature as a learning method for multidimensional data [36, 37] and gave rise to many popular methods such as support vector machine (SVM) [36] and Bayesian Gaussian process models [38]. In the context of association studies, Liu et al. [39] and Liu et al. [40] developed least squares kernel machine and logistic kernel machine regression, respectively. Since its introduction to association studies, there has been substantial work completed in this area in various contexts such as SNP set analysis for genome-wide case control analysis [41], rare variant association testing for sequence data [31], SNP-set analysis for survival outcomes [42], hypothesis testing for a gene effect in presence of complicated interactions [43], marker-set association study with multivariate phenotype [44], and many others.

Simple Kernel Machine Regression To focus our ideas, consider the model

$$Y_i = X_{iC}\beta_C + h(G_{1i}, \ldots, G_{Mi}) + \epsilon_i,$$

where $h(\cdot)$ is an unknown function of the genetic variables. The primary objective is to make statistical inference from this unknown function. The main idea of kernel machine regression is to assume that lies in a reproducing kernel Hilbert space (RKHS) H_S generated by a positive definite kernel function $S(\cdot,\cdot)$. The specification of this kernel function implicitly determines the smoothness property of the unknown function $h(\cdot)$. By Mercer's theorem [45], a kernel function specifies a unique function space spanned by a set of orthogonal basis functions $\{\phi_1(\cdot),\ldots,\phi_J(\cdot)\}$ and one can write $h(G) = \sum_{j=1}^{J}\phi_j(G)\omega_j$, where ω_j's are unknown coefficients. This is called the primal representation of $h(\cdot)$. Equivalently, we can also represent $h(\cdot)$ using the kernel function as $h(G) = \sum_{i=1}^{n} S(G_i,G)\alpha_i$, where α_j are unknown parameters. The later form is called the dual representation. When the argument G is multidimensional, e.g., variables corresponding to markers in a set, it is often more convenient to use the dual representation, since explicit basis functions for primal representation might be complicated to specify, and the number of such functions might be high or even infinite.

In this framework, Liu et al. [39] proposed an estimation procedure for $h(\cdot)$ by maximizing the scaled penalized log-likelihood function

$$-\sum_{i}^{n}\{y_i - X_{iC}\beta_C - h(G_i)\}^2 - \lambda\|h(\cdot)\|^2_{H_S},$$

where $\|h(\cdot)\|^2_{H_S}$ denoted the function norm of $h(\cdot)$, and λ is a smoothing parameter that determines the smoothness of the estimate. By substituting the dual form, the penalized criterion becomes

$$L_{pen} = -\sum_{i}^{n}\left\{y_i - X_{iC}\beta_C - \sum_{j}^{n} S(G_i,G_j)\alpha_j\right\}^2 - \lambda\alpha^T S_G\alpha,$$

where, with a slight abuse of notation, S_G is an $n\times n$ matrix such that $S_{G,ij} = S(G_i,G_j)$. The unknown parameter β_C and α can now be estimated using standard least squares estimation. The tuning parameter can be estimated using GCV or restricted maximum likelihood (REML) estimation [39].

Connection of KMR to GLMM An attractive feature of KMR is its direct connection of GLMM that facilitates the interpretation and computation of this regression methodology. Recall that the parameter estimation is performed by maximizing the penalized criterion L_{pen}. Define $h_G = [h(G_1),\ldots,h(G_n)]$. Then it can be shown that the estimating equations obtained from derivatives of L_{pen} are identical to those of the linear mixed model

$$y = X_C\beta_C + h_G + \epsilon,$$

where one assumes that $\epsilon = (\epsilon_1, \ldots, \epsilon_n)^T \sim N(0, \sigma^2 I_n)$ and the "gene effect" $h_G \sim N(0, \tau S_G)$ with $\tau = \sigma^2 / \lambda$. This connection provides justification for estimating the tuning parameter λ or equivalently τ, and σ^2 using REML. More important, this also enables us to test for no gene effect, that is, $H_0 : h(\cdot) = 0$ by just testing the simpler null hypothesis $H_0 : \tau = 0$. This is achieved by using a score type variance components test for τ [39, 40].

G × E Interaction Testing Using KMR Kernel machine testing is particularly useful when there are multiple environmental variables present, that is, $K_E > 1$. Then one can model the main effects of the marker set variables, and the set of environmental variables, and interactions between these two sets using specific kernel functions. Consider the interaction model

$$Y_i = X_{Ci}^T \beta_C + h_G(G_i) + h_E(E_i) + h_{GE}(G_i, E_i) + \varepsilon_i,$$

where $h_G(\cdot)$ and $h_E(\cdot)$ are unknown functions modeling the main effects of the genetic variables G and the environmental variables E, respectively, and $h_{GE}(\cdot)$ is an unknown function modeling the G × E interaction effect. In the KMR framework, we model $h_z(\cdot)$ by a corresponding kernel function $S_z(\cdot, \cdot)$, $z \in \{G, E, GE\}$.

One needs to be cautious in constructing the interaction kernel $S_{GE}(\cdot, \cdot)$. One natural way to define the interaction kernel is to set $S_{GE}(\cdot, \cdot) = S_G(\cdot, \cdot) S_E(\cdot, \cdot)$, as has been done while studying gene-gene interaction in the KMR framework [46]. However, depending on the kernel used to model the set of environmental variables, it may be not always advisable to create the interaction kernel by the element-wise product of the two kernels used to model the main effects [12] due to the potential risk of duplicate G and E terms in the G × E kernel. Wang et al. [12] proposed a testing methodology for G × E interaction with a carefully built kernel of the interaction effect. For the genetic kernel, Wang et al. proposed using the popular weighted IBS kernel

$$S_G(G_i, G_{i'}) = \frac{1}{2 \sum w_m} \sum_{m=1}^{M} w_m \{2 \times I(G_{mi} = G_{mi'}) + I(|G_{mi} - G_{mi'}| = 1)\},$$

where the weights are taken to be $w_m = q_m^{-3/4}$ [32], where q_m is the minor allele frequency (MAF) of SNP m. The two-way interaction kernel [43] is used to model the set of environmental variables,

$$S_E(E_i, E_{i'}) = 1 + \sum_{\ell=1}^{K_E} X_{E\ell i} X_{E\ell i'} + \sum_{\ell < k} X_{E\ell i} X_{Eki} X_{E\ell i'} X_{Eki'}.$$

It is easy to observe that if one constructs the G × E interaction kernel as simply $S_{GE}((G_i, X_{Ei}), (G_{i'}, X_{Ei'})) = S_G(G_i, G_{i'}) S_E(X_{Ei}, X_{Ei'})$, then the genetic main effects will be duplicated due to the constant in $S_E(\cdot, \cdot)$. Such duplicate terms may cause colinearity in the model that leads to invalid conclusions. An effective yet simple solution of this issue is to redefine the environmental kernel as $S_E^*(X_{Ei}, X_{Ei'}) = S_E(X_{Ei}, X_{Ei'}) - 1$ and then calculate the G × E kernel as $S_{GE}((G_i, X_{Ei}), (G_{i'}, X_{Ei'})) = S_G(G_i, G_{i'}) S_E^*(X_{Ei}, X_{Ei'})$. This approach can also be used for general kernel specification. For example, if we choose to use a polynomial kernel for $S_G(\cdot, \cdot)$ or $S_E(\cdot, \cdot)$ instead of the IBS or interaction kernel, respectively, we can still define $S_z^*(\cdot, \cdot) = K_{z, polynomial}(\cdot, \cdot) - 1$ for $z = G$ or E first, and then construct $S_{GE}(\cdot, \cdot)$ by taking the element-wise product of $S_G(\cdot, \cdot)$ and $S_E(\cdot, \cdot)$.

Following the kernel specification, a score-based test for interaction can be developed using the connection between KMR and GLMM, that is, by considering a linear mixed model representation

$$y = X_C \beta_C + h_G + h_E + h_{GE} + \varepsilon,$$

where $h_z \sim \mathcal{N}(0, \tau_z S_z)$, with $z \in \{G, E, GE\}$. It follows that testing $H_0 : h_{GE}(\cdot) = 0$ is equivalent to testing $H_0 : \tau_{GE} = 0$. Based on these results, the score test for G × E effect can be derived based on the REML likelihood. Specifically, the test statistic for the G × E test is

$$T_{GE} = \frac{1}{2} Y^T P S_{GE} P Y,$$

where $P = V^{-1} - V^{-1} X (X_C^T V^{-1} X_C)^{-1} X_C^T V^{-1}$, and $V = \tau_G S_G + \tau_E S_E + \sigma^2 I_n$. All the quantities used in T_{GE} are computed under the null hypothesis $H_0 : \tau_{GE} = 0$. Wang et al. [13, 12] derived an EM algorithm to perform the estimation under H_0, and also showed that under H_0 that test statistic asymptotically follow a weighted chi-squared distribution. Thus p-values can be obtained by moment matching approaches [35].

Gene-Based G × E for Rare Variants

Many of the introduced approaches can be applied directly for G × E tests with rare variants (i.e., minor allele frequencies <0.05), including SBERIA [10], GESAT [14], SimReg [19, 20], and KMR. PC regressions may be less appropriate for rare variants because variants with lower frequency, and thus less variability, would obtain fewer weights in a PC. Finally, Fan and Lo [47] proposed a model-free approach based on the summation of partitions to evaluate interaction effects for rare variants. However, their method only evaluates the joint effect of G and G × E but not the separate effects.

Gene-Based G × G Interactions

Most of the methods introduced for G × E interactions can also be applied to G × G interaction at the gene level. For example, Wang et al. [48] extended SimReg to study interactions between two genes. Also, several KM regression methods have been proposed to assess gene-level G × G effects. Li and Cui [49] developed a gene-based interaction approach when the quantitative trait is continuous. They used a kernel machine smoothing spline ANOVA approach for this model. Later, Larson and Schaid [46] extended this approach for binary and generalized outcomes. Specifically, Larson and Schaid considered two set of markers G_1 and G_2. The main effects of each set is modeled by kernel matrices K_1 and K_2, respectively, and then the interaction kernel is defined as $K_3 = K_1 \circ K_2$, where $\circ$ denotes the element-wise product of two matrices. Larson and Schaid then consider the mixed effects logistic regression model

$$logit(\mu) = X_C \beta_C + h_1 + h_2 + h_{12},$$

where h_1, h_2 and h_{12} are the main effects of marker sets 1 and 2, and their interaction, respectively. Following the KRM framework, we assume that $h_1 \sim N(0, \tau_1 K_1)$, $h_2 \sim N(0, \tau_2 K_2)$ and $h_{12} \sim N(0, \tau_3 K_{12})$. Thus testing for interaction is equivalent to testing the null hypothesis $H_0 : \tau_3 = 0$. Larson and Schaid proposed a variance components test utilizing the penalized quasi-likelihood approach.

Gene-level G × G tests share a similar principle as the multi-G-multi-E analyses, where both genes being studied for G × G need principal components or variance components to summarize the effect. Therefore, caution and diligence are recommended to avoid the possibility that the interaction kernel is duplicating the main effect if the polynomial kernel or the interactive kernel are used to summarize the gene information, as discussed in Wang et al. [12]. One can still apply the technique of Wang et al [12] to construct the interaction kernel, as described earlier. Recently, Clark et al. [50] reported that they are in the process of developing a KMR approach that attempts to separate the individual effects of the two marker sets from the combined joint effect by projecting the joint kernel onto the individual main effect kernels. The interaction kernel is then built using the residual signal. A score type test can be built a test based on this component, thus avoiding the need to specify any explicit kernel for the interaction.

Summary and Conclusions

Studying complex diseases in the aftermath of genome-wide association studies (GWAS) has led to the development of methods that consider factor sets rather than individual genetic and environmental factors. By pooling information across a set of genetic markers, these methods have improved power, either through aggregating genetic signals or by reducing degrees of freedom, to detect $G \times E$ signals missed by individual SNP analyses. As discussed in this chapter, the marker set approaches to assessing $G \times E$ (and $G \times G$) can be generally categorized into fixed-effects approaches and random-effect approaches. The fixed-effects approaches, e.g., Tukey's one degree-of-freedom test, PC regression, PLS regression, and SBERIA, are easy to apply and are computationally efficient. However, they are subject to inflated type 1 error rates and power loss if the main effect of the genetic factor and environmental factors are not correctly modeled.

The random-effects approaches, which model the main and interaction effects in a semiparametric fashion, would be a more robust choice to model misspecification than fixed-effects methods. The commonly considered random-effect $G \times E$ methods, including GLMM, SimReg, and KMR, can be unified under the framework of GLMM and differ from each other by the kernel function used to summarize genetic similarity and the weights assigned to each variants when quantifying similarity. Random-effects approaches evaluate the $G \times E$ effect by testing the significance of the corresponding single variance component, which is particularly useful when there are sets of genetic variables and environment variables (i.e., $K_E > 1$) to be considered. As we noted, the construction of $G \times E$ kernel for multi-G-multi-E factors requires cautions to avoid double-counting the main effect in the interaction kernel.

The robustness and dimension reduction of random-effects methods do come with a price of computational burden. Because random-effects models use variance components to capture the effects, one needs to estimate nuisance variance components for each effect when assessing $G \times E$ effect. These nuisance variance components are computationally intensive to calculate when the trait of interest is nonquantitative or the sample size is extremely large (e.g., in the order of 10^4). Lin et al. [14] proposed to use a L2 penalty and speed up the computational burden by treating the nuisance variance component as a tuning parameter. We are exploring a fixed-effects approximation to the random-effects model under the null hypothesis to speed up the computational time while enjoying the promises of random-effects approaches [51].

References

1. Kraft, P., Yen, Y.-C., Stram, D. O., Morrison, J., & Gauderman, W. J. (2007). Exploiting gene-environment interaction to detect genetic associations. *Human Heredity, 63*(2), 111–119.

2. Van Os, J., & Rutten, B. P. F. (2009). Gene-environment-wide interaction studies in psychiatry. *American Journal of Psychiatry, 166*(9), 964–966.

3. Murcray, C. E., Lewinger, J. P., & Gauderman, W. J. (2009). Gene-environment interaction in genome-wide association studies. *American Journal of Epidemiology, 169*(2), 219–226.

4. Thomas, D. (2010). Methods for investigating gene-environment interactions in candidate pathway and genome-wide association studies. *Annual Review of Public Health, 31*, 21–36.

5. Zuk, O., Hechter, E., Sunyaev, S. R., & Lander, E. S. (2012). The mystery of missing heritability: Genetic interactions create phantom heritability. *Proceedings of the National Academy of Sciences of the United States of America, 109*(4), 1193–1198.

6. Manolio, T. A., Collins, F. S., Cox, N. J., Goldstein, D. B., Hindorff, L. A., Hunter, D. J., et al. (2009). Finding the missing heritability of complex diseases. *Nature, 461*(7265), 747–753.

7. Hutter, C. M., Mechanic, L. E., Chatterjee, N., Kraft, P., & Gillanders, E. M. (2013). Gene-environment interactions in cancer epidemiology: A National Cancer Institute Think Tank report. *Genetic Epidemiology, 37*(7), 643–657.

8. Chatterjee, N., Kalaylioglu, Z., Moslehi, R., Peters, U., & Wacholder, S. (2006). Powerful multilocus tests of genetic association in the presence of gene-gene and gene-environment interactions. *American Journal of Human Genetics, 79*(6), 1002–1016.

9. Wang, T., Ho, G., Ye, K., Strickler, H., & Elston, R. C. (2009). A partial least-square approach for modeling gene-gene and gene-environment interactions when multiple markers are genotyped. *Genetic Epidemiology, 33*(1), 6–15.

10. Jiao, S., Hsu, L., Bézieau, S., Brenner, H., Chan, A. T., Chang-Claude, J., et al. (2013). SBERIA: set-based gene-environment interaction test for rare and common variants in complex diseases. *Genetic Epidemiology, 37*(5), 452–464.

11. Winham, S. J., & Biernacka, J. M. (2013). Gene-environment interactions in genome-wide association studies: Current approaches and new directions. *Journal of Child Psychology and Psychiatry, and Allied Disciplines, 54*(10), 1120–1134.

12. Wang, Z., Maity, A., Luo, Y., Neely, M. L., & Tzeng, J.-Y. (2015). complete effect-profile assessment in association studies with multiple genetic and environmental factors. *Genetic Epidemiology, 39*(2):122–133.

13. Wang, X., Zhang, D., & Tzeng, J.-Y. (2014). Pathway-guided identification of gene-gene interaction. *Annals of Human Genetics, 78*(6),478–491.

14. Lin, X., Lee, S., Christiani, D. C., & Lin, X. (2013). Test for interactions between a genetic marker set and environment in generalized linear models. *Biostatistics, 14*(4), 667–681.

15. Voorman, A., Lumley, T., McKnight, B., & Rice, K. (2011). Behavior of QQ-plots and genomic control in studies of gene-environment interaction. *PLoS One, 6*(5), e19416.

16. Lin, X. (1997). Variance component testing in generalised linear models with random effects. *Biometrika, 84*(2), 309–326.

17. O'sullivan, F., Yandell, B. S., & Jr, W. J. R. (1986). Automatic smoothing of regression functions in generalized linear models. *Journal of the American Statistical Association, 81*(393), 96–103.

18. Davies, R. B. (1980). The distribution of a linear combination of chi-square random variables. *Applied Statistics, 29*(3): 323–333.

19. Tzeng, J.-Y., Zhang, D., Pongpanich, M., Smith, C., McCarthy, M. I., Sale, M. M., et al. (2011). Studying gene and gene-environment effects of uncommon and common variants on continuous traits: A marker-set approach using gene-trait similarity regression. *American Journal of Human Genetics, 89*(2), 277–288.

20. Zhao, G., Marceau, R., Zhang, Z., & Tzeng, J.-Y. (2015). Assessing gene-environment interactions for common and rare variants with binary traits using gene-trait similarity regression. *Genetics, 199*(3):695–710.

21. Haseman, J. K., & Elston, R. C. (1972). The investigation of linkage between a quantitative trait and a marker locus. *Behavior Genetics, 2*(1), 3–19.

22. Elston, R. C., Buxbaum, S., Jacobs, K. B., & Olson, J. M. (2000). Haseman and Elston revisited. *Genetic Epidemiology, 19*(1), 1–17.

23. Tzeng, J.-Y., Devlin, B., Wasserman, L., & Roeder, K. (2003). On the identification of disease mutations by the analysis of haplotype similarity and goodness of fit. *American Journal of Human Genetics, 72*(4), 891–902.

24. Beckmann, L., Fischer, C., Obreiter, M., Rabes, M., & Chang-Claude, J. (2005). Haplotype-sharing analysis using Mantel statistics for combined genetic effects. BMC Genetics, 6 Suppl 1(Suppl 1), S70.

25. Tzeng, J.-Y., Zhang, D., Chang, S.-M., Thomas, D. C., & Davidian, M. (2009). Gene-trait similarity regression for multimarker-based association analysis. Biometrics, 65(3), 822–832.

26. Tzeng, J.-Y., Lu, W., & Hsu, F.-C. (2014). Gene-level pharmacogenetic analysis on survival outcomes using gene-trait similarity regression. Annals of Applied Statistics, 8(2), 1232–1255.

27. Schaid, D. J. (2010). Genomic similarity and kernel methods I: Advancements by building on mathematical and statistical foundations. Human Heredity, 70(2), 109–131.

28. Wessel, J., & Schork, N. J. (2006). Generalized genomic distance-based regression methodology for multilocus association analysis. American Journal of Human Genetics, 79(5), 792–806.

29. Schaid, D. J. (2010). Genomic similarity and kernel methods II: Methods for genomic information. Human Heredity, 70(2), 132–140.

30. Price, A. L., Kryukov, G. V., de Bakker, P. I. W., Purcell, S. M., Staples, J., Wei, L.-J., et al. (2010). Pooled association tests for rare variants in exon-resequencing studies. American Journal of Human Genetics, 86, 832–838.

31. Wu, M. C., Lee, S., Cai, T., Li, Y., Boehnke, M., & Lin, X. (2011). Rare-variant association testing for sequencing data with the sequence kernel association test. American Journal of Human Genetics, 89(1), 82–93.

32. Pongpanich, M., Neely, M. L., & Tzeng, J.-Y. (2012). On the aggregation of multimarker information for marker-set and sequencing data analysis: Genotype collapsing vs. similarity collapsing. Frontiers in Genetics, 2, 110.

33. Zhang, D., & Lin, X. (2003). Hypothesis testing in semiparametric additive mixed models. Biostatistics, 4(1), 57–74.

34. Tzeng, J.-Y., & Zhang, D. (2007). Haplotype-based association analysis via variance-components score test. American Journal of Human Genetics, 81(5), 927–938.

35. Duchesne, P., & Lafaye De Micheaux, P. (2010). Computing the distribution of quadratic forms: Further comparisons between the Liu-Tang-Zhang approximation and exact methods. Computational Statistics & Data Analysis, 54(4), 858–862.

36. Vapnik, V. (1998). Statistical Learning Theory. Wiley.

37. Schölkopf, B., & Smola, A. J. (2002). Learning with Kernels: Support Vector Machines, Regularization, Optimization, and Beyond. MIT Press.

38. Rasmussen, C. E. (2006). Gaussian Processes for Machine Learning. MIT Press.

39. Liu, D., Lin, X., & Ghosh, D. (2007). Semiparametric regression of multi-dimensional genetic pathway data: Least squares kernel machines and linear mixed models. *Biometrics*, *63*(4): 1079–1088.

40. Liu, D., Ghosh, D., & Lin, X. (2008). Estimation and testing for the effect of a genetic pathway on a disease outcome using logistic kernel machine regression via logistic mixed models. *BMC Bioinformatics*, *9*(1), 292.

41. Wu, M. C., Kraft, P., Epstein, M. P., Taylor, D. M., Chanock, S. J., Hunter, D. J., et al. (2010). Powerful SNP-set analysis for case-control genome-wide association studies. *American Journal of Human Genetics*, *86*(6), 929–942.

42. Lin, X., Cai, T., Wu, M. C., Zhou, Q., Liu, G., Christiani, D. C., et al. (2011). Kernel machine SNP-set analysis for censored survival outcomes in genome-wide association studies. *Genetic Epidemiology*, *35*(7), 620–631.

43. Maity, A., & Lin, X. (2011). Powerful tests for detecting a gene effect in the presence of possible gene-gene interactions using garrote kernel machines. *Biometrics*, *67*(4), 1271–1284.

44. Maity, A., Sullivan, P. F., & Tzeng, J.-Y. (2012). Multivariate phenotype association analysis by marker-set kernel machine regression. *Genetic Epidemiology*, *36*(7), 686–695.

45. Cristianini, N., & Shawe-Taylor, J. (2000). *An Introduction to Support Vector Machines and Other Kernel-based Learning Methods*. Cambridge University Press.

46. Larson, N. B., & Schaid, D. J. (2013). A kernel regression approach to gene-gene interaction detection for case-control studies. *Genetic Epidemiology*, *37*(7), 695–703.

47. Fan, R., & Lo, S.-H. (2013). A robust model-free approach for rare variants association studies incorporating gene-gene and gene-environmental interactions. *PLoS One*, *9*(5), e98083.

48. Wang, X., Epstein, M. P., & Tzeng, J.-Y. (2014). Analysis of gene-gene interactions using gene-trait similarity regression. *Human Heredity*, *78*(1), 17–26.

49. Li, S., & Cui, Y. (2012). Gene-centric gene–gene interaction: A model-based kernel machine method. *Annals of Applied Statistics*, *6*(3), 1134–1161.

50. Clark, J. J., Maity, A., Harmon, Q., Engel, S. M., Epstein, M. P., & Wu, M. C. (submitted). Gene and region based testing of gene-gene interactions for quantitative traits with the snp-set kernel interaction test.

51. Marceau, R., Lu, W., Holloway, S., Sale, M.M., Worrall, B.B., Williams, S.R., Hsu, F., & Tzeng, J.-Y. (2015). A fast multiple-kernel method with applications to detect gene-environment interaction. *Genetic Epidemiology*, *39*(6), 456–468.

4 Set-Based Gene × Environment Interaction Tests for Complex Diseases with Application to Genome-Wide Association and Sequencing Studies

Shuo Jiao

Both genetic (G) and environmental (E) factors impact common complex diseases, such as cancer, diabetes, or cardiovascular diseases. For most of these diseases, several environmental factors and a rapidly increasing number of genetic factors have been identified [1]. However, little is understood about the interplay between G and E. Some exceptions include observed interactions between smoking and the *GSTM1* deletion and a tag SNP in *NAT2* in bladder cancer [2, 3], *ADH7* variants and alcohol consumption in upper aerodigestive cancers [4], and *GRIN2A* variants and coffee consumption in Parkinson's disease [5].

Several potential factors have contributed to the limited numbers of confirmed gene-environment interactions (G × E). For example, measurement error and data harmonization issues across studies for the environmental factors could dilute the interaction signals. Probably more important, the statistical power to detect an interaction is much smaller compared to detecting a main effect. In fact, it has been shown that the detection of an interaction needs at least approximately four times as many subjects as are needed to detect a main genetic effect of comparable effect size [6].

A number of methods have been proposed to enhance the power of detecting G × E interactions and include the case-only test [7, 8], the empirical Bayes method [9], and the Bayesian Model Averaging method [10]. Two types of screening methods have also been proposed to reduce the multiple testing burden in genome-wide G × E search: correlation based screening [11] and marginal association based screening [12]. Toward this end, several recent methods were developed to combine and maximize the advantages of different screening and testing techniques, such as the hybrid method by Murcray et al. (2011) [13] and the cocktail method by Hsu et al. (2012) [14].

Those efforts focus on improving the power of detecting $G \times E$ interactions for individual markers. On the other hand, a set-based method that includes multiple rather than individual markers not only can enhance the statistical power by aggregating multiple signals in the same set, but also can greatly reduce the number of tests to be performed and thus reduce the multiple testing burden. Most of the existing set-based methods are for detecting genetic main effects, which means testing the association between a set of SNPs and a phenotype. Tzeng et al. (2011) provided a summary of those methods, which included burden tests that compute the weighted sum of genotypes across markers [15–18], methods that exploit the pairwise genetic similarity among samples [19–26], variance component methods [27–32], a method that combines p-values within a gene [33], group additive regression [34], Tukey's model [35], and an entropy-based method [36]. Set-based methods have drawn more attention in the sequencing studies because of the rarity of the variants; for example, several variations of the burden tests [37–43] and variance component tests [32, 44] have been proposed for sequencing data.

In contrast, few methods have been proposed for set-based $G \times E$ tests. Tzeng et al. (2011) developed a method to test for interaction between a set of markers and an environment variable by extending the set-based genetic similarity method to the $G \times E$ setting [45]. Because there is no competing method, they compared the new method with the benchmark minimum p-value method and their method showed favorable performance. However, their method was designed for a continuous outcome and cannot be applied to a case-control study for complex diseases.

In the next section, we introduce two existing set-based $G \times E$ tests that can be applied to complex diseases and compare their performances using extensive simulations, in both GWAS and rare variant settings. After that, we discuss several practical issues and potential improvements for existing set-based $G \times E$ methods.

Methods

Notations and Models
Suppose there are N subjects and the disease status is denoted by D_i (= 0 or 1) for subject i, $i = 1,...N$. Assume E_i is the environmental variable, $\mathbf{X}_i = (X_{i1},...X_{iq})$ is a vector of q potential confounder covariates, and

$G_i = (G_{i1}, ...G_{ip})$ is a vector of p genetic markers. The interaction model between the set of p markers and the environmental variable is:

$$logit\{P(D_i = 1)\} = \alpha_0 + \alpha_1 E_i + G_i\alpha_2 + X_i\alpha_3 + E_iG_i\beta, \tag{4.1}$$

where logit() is the logit link function; α_0 is the intercept; α_1 is the coefficient for the main effect of E_i; α_2 is the p × 1 vector of coefficients for G_i; α_3 is the q × 1 vector of coefficients for X_i; $E_iG_i = (E_iG_{i1}, ...E_iG_{ip})$; $\beta = (\beta_1, ..., \beta_p)^T$ is the p × 1 vector of interaction coefficients. The null hypothesis for interaction effects is $H_0 : \beta = 0$.

A natural approach to developing a set-based G × E test is directly extending the set-based main effect test by treating the interaction term (usually the product of G and E) as a new genetic variable. For example, the existing burden test computes the (un)weighted sum of the genotypes (minor alleles counts) across SNPs in the set and tests whether the sum is associated with the phenotype. A simple extension of burden tests to the G × E setting would be to sum the interaction terms (products) of G and E instead of summing over the G's alone. However, assumptions that are reasonable for main effects may not be reasonable for G × E, i.e., the power of burden tests for rare variants depends on the assumption that most rare missense variants are deleterious, but it is not reasonable to assume all G × E's have the same direction. To solve the problem, so far two set-based G × E methods that can be applied to complex diseases have been proposed.

SBERIA

It can be noted that this naive extension of burden test to G × E fails to exploit some unique characteristics of G × E. For instance, one major difficulty in the set-based test is the lack of prior information on which SNPs are null and what directions the effects are. In the set-based main effect tests, there have been several attempts trying to solve this issue. Han and Pan (2010) used the signs of the marginal effect to determine the direction of the main effect [42]. Lin and Tang (2011) used the corresponding regression coefficient plus a constant as the weight for each marker [46]. Cai et al. (2012) proposed to weight each marker based on the z-score of its effect [47]. A common characteristic of these methods is that the statistics used to weight the markers are not independent of the main effect test. Hence, permutation is needed to estimate the null distribution and maintain the correct type 1 error, which is computationally intensive. In contrast, for

G × E, there are screening statistics that are informative for weighting the markers but still independent of the interaction test. Therefore, it would be appealing to take advantage of this desirable feature of the G × E test.

Correlation screening has been established as an efficient screening tool for the G × E test [11]. Consider the following simple example to understand the rationale of the correlation screening. Suppose there is a rare disease D, an environmental variable E (= 0 or 1), and a genetic variable G (= 0 or 1). G and E are assumed to be independent in controls (and because of the rarity of the disease, also approximately independent in general population). Assume there is a positive interaction between E and G such that the disease risk would only increase when both E = 1 and G = 1. Then we expect to see more E = 1 & G = 1 combinations in the cases, which means G and E will be positively correlated in the cases. Because G and E are independent in controls, they will be also positively correlated in the combined case-control samples. From this simple example, we can see that the correlation between G and E combined case-control samples can be useful as a screening statistic for interaction between G and E. In addition, the direction of the correlation can inform the direction of the interaction. More important, it has been shown both by Murcray et al. [11] and Dai et al. [48] that the correlation screening in combined case-control samples is asymptotically independent of the G × E test, no matter whether G and E are independent or not. Motivated by this, Jiao et al. (2013) proposed a set-based gene environment interaction test (SBERIA) [49].

First, the correlation between E_i and G_{ij} (j = 1 to p) in (1) was computed by either fitting a logistic regression (when E_i is binary) or a linear regression (when E_i is continuous) with E_i as the response and G_{ij} as the predictor. Then each SNP j (j = 1 to p) has a Z-score Z_j for the correlation between E_i and G_{ij}. Then the following logistic regression model is fitted:

$$logit\{P(D_i = 1)\} = \alpha_0 + \alpha_1 E_i + \mathbf{G_i}\boldsymbol{\alpha}_2 + \mathbf{X_i}\boldsymbol{\alpha}_3 + \rho E_i \mathbf{G_i}\hat{\mathbf{w}}, \tag{4.2}$$

where $\hat{\mathbf{w}} = (\hat{w}_1,...\hat{w}_p)^T$ is the weight vector and $\hat{w}_j = I(|Z_j| > \theta_N)sign(Z_j) + \varepsilon$. $I(x)$ is an indicator function that equals 0 when x is false and 1 when x is true. $sign(x) = 1$ when x > 0, –1 when x < 0 and 0 when x = 0. $\theta_N = o(N^{1/2})$ and ε are prespecified positive constants. The hypothesis of interest is $H_0 : \rho = 0$.

As we can see, $E_i \mathbf{G_i}\hat{\mathbf{w}}$ is the weighted sum of the interaction terms and the weight, which can be 1, –1, or 0 (if we ignore ε), is determined by correlation Z-score Z_j. $|Z_j|$ measures the strength of the correlation signal so $I(|Z_j| > \theta_N)$ selects only markers showing correlation signals that are greater than a threshold. $\theta_N = o(N^{1/2})$ because $I(|Z_j| > \theta_N)$ is expected to converge to 0 as $N \to \infty$ when there is no correlation between G and E in the combined sample and converge to 1 when there is correlation. For the selected marker (markers with $I(|Z_j| > \theta_N) = 1$), the direction of the interaction term is determined by the direction of the correlation ($sign(Z_j)$). This is inspired by the observation that the directions of interaction and correlation tend to correspond with one another in the previous simple example. The addition of a constant ε ensures that a weight will be assigned if no marker is selected.

In addition to correlation screening, marginal association screening [12] can also be used to inform the G × E test and similar as the correlation screening, marginal screening is also independent of the interaction test. Therefore, correlation and marginal screenings can be combined to make SBERIA more robust. Specifically, instead of using $\hat{w}_j = I(|Z_j| > \theta_N)sign(Z_j) + \varepsilon$ in (2), define

$$\hat{w}_j = I(|C_j| > \tau_N)sign(C_j) + \varepsilon, \tag{4.3}$$

where $C_j = Z_j$ if $|Z_j| > |M_j|$; otherwise $C_j = M_j$ (M_j is the wald statistic of the marginal association for marker j); τ_N is also defined such that $\text{Prob}(|C_j| > \tau_N) = 0.1$ under the null. θ_N and ε need to be specified for SBERIA. In Jiao et al. (2013), θ_N was set to a constant such that $\text{Prob}(|Z_j| > \theta_N) \approx 0.1$ under the null. ε was set to a very small value (0.0001) so that it does not affect the weight if $I(|Z_j| > \theta_N)sign(Z_j)$ is not 0.

In summary, SBERIA first selects markers of which the screening signal strength is greater than a threshold. For the selected markers, a weighted sum of their interaction terms was computed where the weight = 1 if the corresponding screening statistic is positive and –1 otherwise. As the screening statistic is independent of the interaction test, regular logistic regression can be used to test the hypothesis without requiring permutation.

GESAT

In a set-based G × E test, it is not reasonable to assume that all the variants in the set are causal and all the interaction effects are in the same direction. SBERIA tackles this problem by taking advantage of the screening statistics for the interaction test. On the other hand, variance component tests allow for heterogeneous effects within a set and therefore are suitable for testing interaction effects. One of the most popular variance component tests is the SNP-set Kernel Association Test (SKAT), which has been proposed to test for association between a set of variants and phenotypes [44]. Lin et al. (2013) proposed a set-based G × E test called GESAT by extending the SKAT to the G × E setting [50]. GESAT uses SKAT with the linear kernel to test for the G × E effects while incorporating the main SNP effects as covariates. It was observed that if the number of variants in a set is large, the main effect part of the model can cause convergence issues; therefore, GESAT uses ridge regression when fitting main effects.

Comparison between SBERIA and GESAT

Both SBERIA and GESAT have been shown to keep the correct type 1 error and outperform several benchmark methods such as the likelihood ratio test and minimal p-value test by a large margin [49, 50]. However, so far there has been no comparison between SBERIA and GESAT. In this section, we conducted extensive simulation to evaluate the performance of those two methods.

GWAS Setting We mimicked the real GWAS data by generating a set of markers based on the realistic LD structure within the *SMAD7* gene. *SMAD7*, short for *SMAD* family member 7, is a gene located at 18q21.1. It is known to interact with the TGF-beta receptor and several SNPs in this region have been found to associate significantly with colorectal cancer risk [51–54]. *SMAD* spans from 44,700k bp to 44,731k bp and has 48 SNPs from Hapmap II release 24 [55], which is close to the median number (= 43) of SNPs per gene [56]. Out of the 48 SNPs, 21 were genotyped in Illumina Human1M. We extracted the haplotypes of the 21 SNPs from the phased Hapmap data and randomly paired haplotypes such that the simulated marker set maintains the same LD structure as the 21 SNPs in the Hapmap. We chose two SNPs rs4939827 and rs7351039 from the 21 SNPs and make them the hidden causal SNPs in the simulation. The two SNPs were chosen such that one is common (rs4939827, MAF = 0.49) and one is less common (rs7351039, MAF = 0.08). The two chosen SNPs are not in LD with each other and both SNPs were tagged by some other SNPs. The other 19 SNPs were considered as the marker set in the simulation.

The disease status was generated based on the following model:

$$logit\{P(D_i = 1)\} = \alpha_0 + \gamma E_i + \alpha_1 G_{i1} + \alpha_2 G_{i2} + \beta_1 E_i G_{i1} + \beta_2 E_i G_{i2}, \tag{4.4}$$

where $\alpha_0 = \exp(-5)$, representing a relatively rare disease. G_{i1} and G_{i2} are the simulated genotypes (=0,1,or 2) for rs4939827 and rs7351039, respectively. E_i is the environmental variable. We tried two ways of generating E_i: 1) E_i is continuous: $E_i \sim N(0,1)$; 2) E_i is binary: $E_i \sim$ Bernoulli$(p = 0.3)$.

To evaluate the power, we set $\beta_1 = \log(1.05)$, $\log(1.10)$, $\log(1.15)$, $\log(1.20)$, $\log(1.25)$, or $\log(1.3)$ when E_i is continuous and $\beta_1 = \log(1.1)$, $\log(1.2)$, $\log(1.3)$, $\log(1.4)$, $\log(1.5)$ or $\log(1.6)$ when E_i is binary. The values of β_1 were chosen such that the power was in a reasonable range. For each value of β_1, β_2 can take three values β_1, $-\beta_1$, or 0, which represents situations where two signals are in the same direction, in the different direction or when there is only one signal, respectively. The main effects α_1 and α_2 were set to 0. We also tried other values for the main effects and the results were quantitatively similar. As before, we randomly generated 1,000 cases and 1,000 controls. We evaluated SBERIA and GESAT for the power performance. Each parameter setting for the simulation was repeated 2,000 times, and we used a significance level of 0.05. The results were summarized in figure 4.1. In general, SBERIA and GESAT have similar power across various simulation scenarios. One interesting observation is that SBERIA tends to have better power than GESAT when effect sizes are larger and GESAT seems to perform better under smaller effect sizes. This makes sense because when effect sizes are large, the screening statistics for SBERIA are more informative. In addition, the performance of SBERIA is better when E is binary.

Rare Variant Setting The increasing popularity of set-based methods is driven mainly by the need for power when testing the associations between rare variants from sequencing studies and outcomes of interest. Thus, in this section we focus on evaluating the performance of SEBRIA and GESAT in the rare variant setting.

In the simulation, the disease status was generated based on the following model:

$$logit\{P(D_i = 1)\} = \alpha_0 + \gamma E_i + \sum_{j=1}^{p} \alpha_j G_{ij} + \sum_{j=1}^{p} \beta_j E_i G_{ij}, i = 1, \ldots n; \tag{5.5}$$

where $\alpha_0 = \exp(-5)$, denoting a relatively rare disease; $\gamma = \log(1.2)$ is the effect size for the environmental factor E_i; E_i is assumed to be a binary

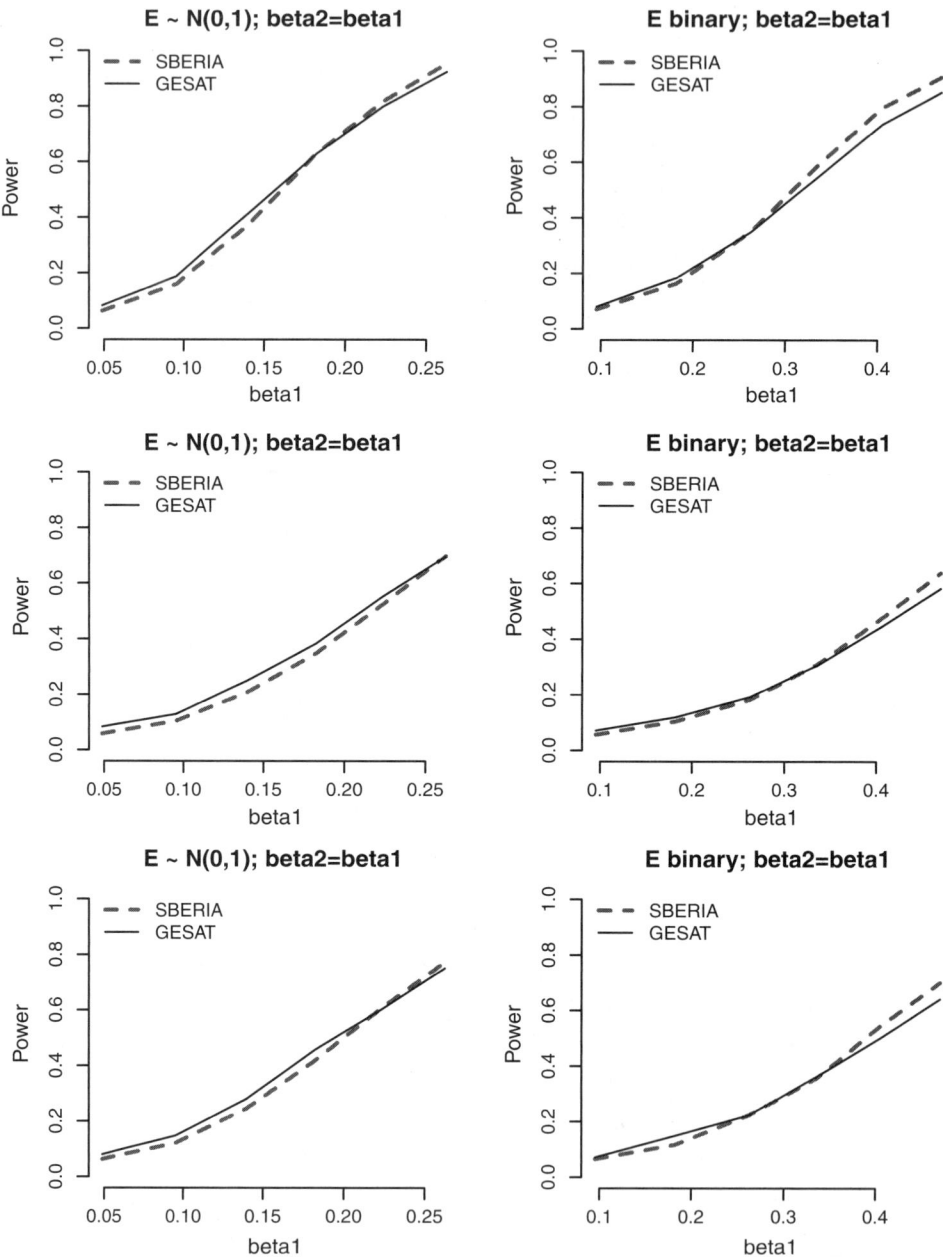

Figure 4.1
Power comparison between SBERIA and GESAT in GWAS settings. The three plots on the left are results when E_i was generated as continuous variable and the plots on the right are for binary E_i's. The top plots are for simulation scenarios where $\beta_1 = \beta_2$; the plots in the middle are for scenarios where $\beta_1 = -\beta_2$; the bottom plots are for scenarios where $\beta_2 = 0$.

variable with frequency 0.3; p is the number of variants in the set; G_{ij} is the genotype of variant j in sample i; α_j's and β_j's are the main effects and interaction effects, respectively. For each of the simulations, we generated a large population based on model (4.6) and randomly selected 2,000 cases and 2,000 controls. For each scenario, we generated 2,000 simulated datasets.

We used two models to generate datasets for the evaluation of power.

1. Model 4.1: p (=10, 20, 40) variants G_{ij} (j=1 to p) were generated with $MAF_j \sim$ Uniform (0.001, 0.05) under Hardy-Weinberg equilibrium. E_i was generated as a binary variable Bernoulli(0.3). We set a background main effect for each variant as $\alpha_j \sim Normal(0, \log(1.5)/2)$. The interaction effects β_j's in (6) were generated as $c^* |\log 10(MAF_j)|$ $* Bernoulli(P_{causal}) * \{1-2* Bernoulli(P_{negative})\}$ where P_{causal}=0.2, 0.5, 0.8 and $P_{negative}$=0.2, 0.5. In other words, every variant has probability P_{causal} of being causal (with interaction effect), and if causal, the interaction effect has probability $P_{negative}$ of being negative. In addition, the rarer variants have larger effect sizes [44]. For the variants with an interaction effect, the main effects were set as 0, representing a synergistic interaction model. In order to see differences among methods, c was chosen such that the resulting power was not too high or too low. It can be different for different scenarios, and thus the actual power is not comparable across scenarios; for each scenario, though, the methods are directly comparable.

2. Model 4.2: The same simulation settings were used as in model 4.1 except that for the variants with interaction effects ($\beta_j \neq 0$), the corresponding main effects were set to be $-0.5\beta_j$, which represents a qualitative interaction model because the main effect is in opposite direction to the interaction effect.

For each model, SBERIA and GESAT were applied to perform the set-based G × E test, and their power was estimated based on significance level 0.05.

The power comparison results under model 4.1 were summarized in figure 4.2. SBERIA generally is more powerful than GESAT under this simulation setting. The advantage of SBERIA is more obvious when the percent of casual variants is large. On the other hand, when the interaction model is qualitative (model 4.2), figure 4.3 shows that GESAT is always more

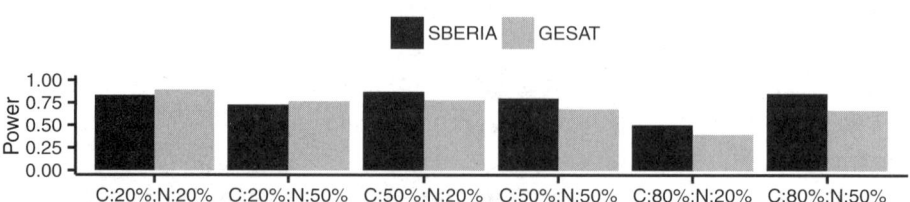

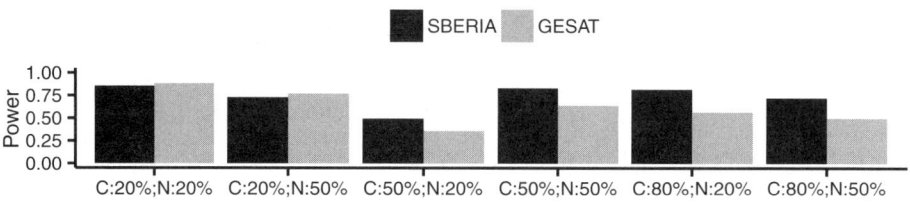

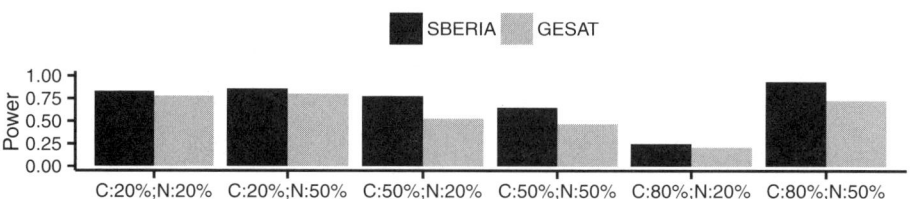

Figure 4.2
Power of SBERIA and GESAT under model 4.1. Different proportion of causal variants ($C=P_{causal}$) and proportion of causal variants with negative effects ($N=P_{negative}$) were used.

powerful, which indicates that the screening statistics in SBERIA do not work well under the qualitative interaction model.

Practical Issues and Potential Improvements

Practical Issues: The Choice of θ_N for SBERIA

Recall that for SBERIA, θ_N is a parameter that needs to be determined by the user. The recommended θ_N corresponds to the p-value cutoff 0.1. In this section, we try different cutoff values to see if they have any effect on

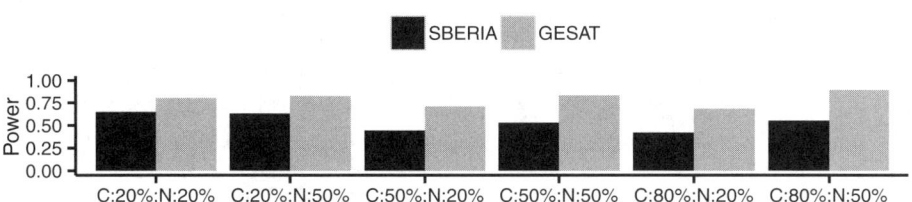

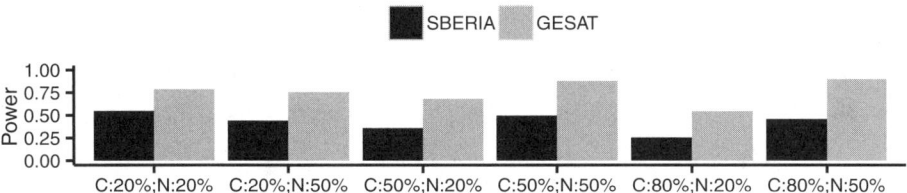

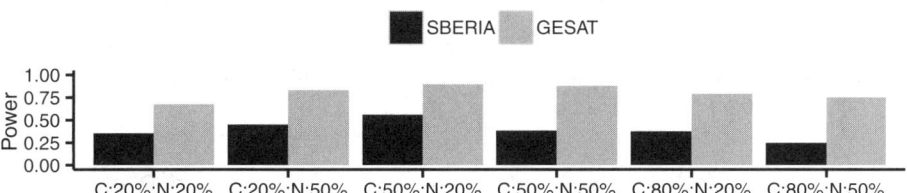

Figure 4.3
Power of SBERIA and GESAT under model 4.2. Different proportion of causal variants (C=P_{causal}) and proportion of causal variants with negative effects (N=$P_{negative}$) were used.

the power. The same simulation setting as in the preceding section is used, except that we try three different θ_N's that correspond to p-value cutoffs 0.05, 0.1, and 0.2. The results are summarized in figure 4.4. We can see that the power of SBERIA does not change substantially under different θ_N's in the simulation.

Potential Improvements

There are several possible improvements that can be made to SBERIA and GESAT. First, for SBERIA, θ_N needs to be chosen such that it corresponds

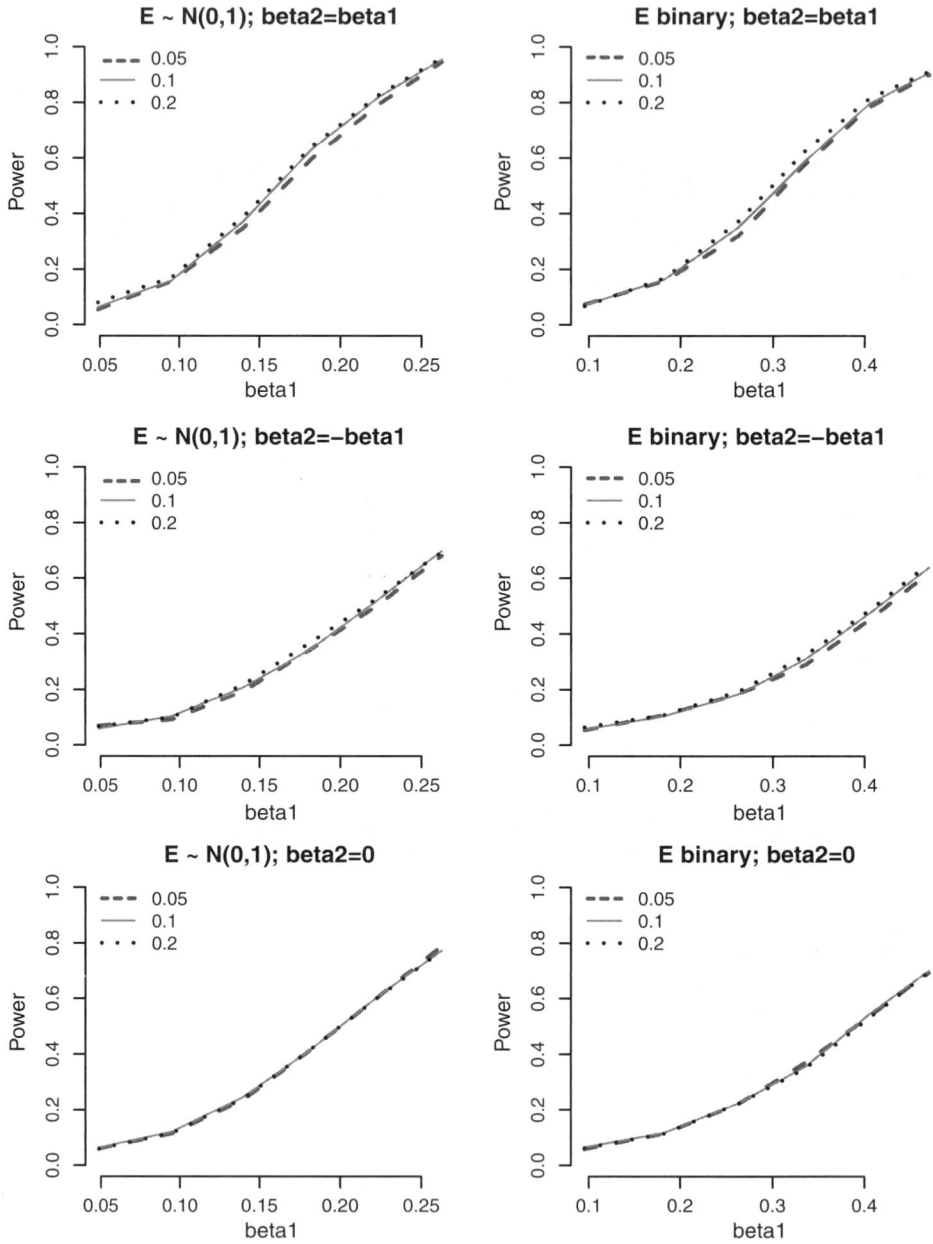

Figure 4.4

Power comparison between SBERIA under different θ_N's in the GWAS settings. The three plots on the left result when E_i was generated as a continuous variable, and the plots on the right are for binary E_i's. The top plots are for simulation scenarios where $\beta_1 = \beta_2$; the plots in the middle are for scenarios where $\beta_1 = -\beta_2$; the bottom plots are for scenarios where $\beta_2 = 0$.

to p-value cutoff 0.1. In the previous section, we tried other p-value cutoffs such as 0.05 and 0.2 in the simulation, and the power of SBERIA does not change substantially. However, it should be noted that the minor allele frequency affects the power of the screening, and the SNPs with larger MAF will be more likely to pass the screening compared to less common SNPs. Hence, it is desired to select a threshold varied with MAF, which is very difficult in practice. In addition, the current weighting of SBERIA is either 1, –1, or 0. Further work should explore whether the use of more advanced weight, such as the effect size or the test statistics of the correlation screening, would increase power. Furthermore, more sophisticated methods can be built upon the framework of our method. For example, SBERIA drops markers that are not selected based on screening. However, as the screening is not perfect, those SNPs may still contain useful information. Hence, it could potentially increase power to apply the traditional method (i.e., variance component based method) to the unselected SNPs and combine the results from the selected and unselected SNPs. Similar improvements can be done to GESAT. GESAT is an extension of SKAT. On the other hand, SKAT-O [57] combines SKAT test and burden test and has been shown to be more robust. Therefore, it is natural to also extend SKAT-O to the G × E setting.

In summary, while both SBERIA and GESAT show promising performances over benchmark methods, the power comparison between the two methods depends on the underlying interactions. SBERIA tends to perform better in synergistic interaction models and when the interaction effects are relatively large. GESAT performs better under qualitative interaction models and when the interaction effects are small. In addition, there are several improvements worth pursuing, which have the potential to further boost the power and robustness of the two methods.

References

1. Hindorff, L. A., Sethupathy, P., Junkins, H. A., Ramos, E. M., Mehta, J. P., Collins, F. S., et al. (2009). Potential etiologic and functional implications of genome-wide association loci for human diseases and traits. *Proceedings of the National Academy of Sciences of the United States of America, 106,* 9362–9367. doi:.10.1073/pnas.0903103106

2. Rothman, N., Garcia-Closas, M., Chatterjee, N., Malats, N., Wu, X., Figueroa, J. D., et al. (2010). A multi-stage genome-wide association study of bladder cancer identifies multiple susceptibility loci. *Nature Genetics, 42,* 978–984. doi:.10.1038/ng.687

3. García-Closas, M., Malats, N., Silverman, D., Dosemeci, M., Kogevinas, M., Hein, D. W., et al. (2005). NAT2 slow acetylation, GSTM1 null genotype, and risk of bladder cancer: Results from the Spanish Bladder Cancer Study and meta-analyses. *Lancet*, *366*, 649–659. doi:.10.1016/S0140-6736(05)67137-1

4. Hashibe, M., McKay, J. D., Curado, M. P., Oliveira, J. C., Koifman, S., Koifman, R., et al. (2008). Multiple ADH genes are associated with upper aerodigestive cancers. *Nature Genetics*, *40*, 707–709. doi:.10.1038/ng.151

5. Hamza, T. H., Chen, H., Hill-Burns, E. M., Rhodes, S. L., Montimurro, J., Kay, D. M., et al. (2011). Genome-wide gene-environment study identifies glutamate receptor gene GRIN2A as a Parkinson's disease modifier gene via interaction with coffee. *PLOS Genetics*, *7*, e1002237. doi:.10.1371/journal.pgen.1002237

6. Smith, P. G., & Day, N. E. (1984). The design of case-control studies: The influence of confounding and interaction effects. *International Journal of Epidemiology*, *13*, 356–365. http://www.ncbi.nlm.nih.gov/pubmed/6386716 accessed July 15, 2012.

7. Piegorsch, W. W., Weinberg, C. R., & Taylor, J. A. (1994). Non-hierarchical logistic models and case-only designs for assessing susceptibility in population-based case-control studies. *Statistics in Medicine*, *13*, 153–162. http://www.ncbi.nlm.nih.gov/pubmed/8122051 accessed September 18, 2012.

8. Chatterjee, N., & Carroll, R. J. (2005). Semiparametric maximum likelihood estimation exploiting gene-environment independence in case-control studies. *Biometrika*, *92*, 399–418. doi:.10.1093/biomet/92.2.399

9. Mukherjee, B., & Chatterjee, N. (2008). Exploiting gene-environment independence for analysis of case-control studies: An empirical Bayes-type shrinkage estimator to trade-off between bias and efficiency. *Biometrics*, *64*, 685–694. doi:.10.1111/j.1541-0420.2007.00953.x

10. Li, D., & Conti, D. V. (2009). Detecting gene-environment interactions using a combined case-only and case-control approach. *American Journal of Epidemiology*, *169*, 497–504. doi:.10.1093/aje/kwn339

11. Murcray, C. E., Lewinger, J. P., & Gauderman, W. J. (2009). Gene-environment interaction in genome-wide association studies. *American Journal of Epidemiology*, *169*, 219–226. doi:.10.1093/aje/kwn353

12. Kooperberg, C., & Leblanc, M. (2008). Increasing the power of identifying gene × gene interactions in genome-wide association studies. *Genetic Epidemiology*, *32*, 255–263. doi:.10.1002/gepi.20300

13. Murcray, C. E., Lewinger, J. P., Conti, D. V., Thomas, D. C., & Gauderman, W. J. (2011). Sample size requirements to detect gene-environment interactions in genome-wide association studies. *Genetic Epidemiology*, *35*, 201–210. doi:.10.1002/gepi.20569

14. Hsu, L., Jiao, S., Dai, J. Y., Hutter, C., Peters, U., et al. (2012). Powerful cocktail methods for detecting genome-wide gene-environment interaction. *Genetic Epidemiology*, *36*, 183–194.

15. Li, M., Wang, K., Grant, S. F. A., Hakonarson, H., & Li, C. (2009). ATOM: A powerful gene-based association test by combining optimally weighted markers. *Bioinformatics*, *25*, 497–503. doi:.10.1093/bioinformatics/btn641

16. Wang, T., & Elston, R. C. (2007). Improved power by use of a weighted score test for linkage disequilibrium mapping. *American Journal of Human Genetics*, *80*, 353–360. doi:.10.1086/511312

17. Gauderman, W. J., Murcray, C., Gilliland, F., & Conti, D. V. (2007). Testing association between disease and multiple SNPs in a candidate gene. *Genetic Epidemiology*, *31*, 383–395. doi:.10.1002/gepi.20219

18. Wang, K., & Abbott, D. (2008). A principal components regression approach to multilocus genetic association studies. *Genetic Epidemiology*, *32*, 108–118. doi:.10.1002/gepi.20266

19. Tzeng, J.-Y., Devlin, B., Wasserman, L., & Roeder, K. (2003). On the identification of disease mutations by the analysis of haplotype similarity and goodness of fit. *American Journal of Human Genetics*, *72*, 891–902. doi:.10.1086/373881

20. Schaid, D. J., McDonnell, S. K., Hebbring, S. J., Cunningham, J. M., & Thibodeau, S. N. (2005). Nonparametric tests of association of multiple genes with human disease. *American Journal of Human Genetics*, *76*, 780–793. doi:.10.1086/429838

21. Dempfle, A., Hein, R., Beckmann, L., Scherag, A., Nguyen, T. T., Schäfer, H., et al. (2007). Comparison of the power of haplotype-based versus single- and multilocus association methods for gene × environment (gene × sex) interactions and application to gene × smoking and gene × sex interactions in rheumatoid arthritis. *BMC Proceedings*, *1*(Suppl 1), S73. http://www.biomedcentral.com/1753-6561/1/S1/S73 accessed September 24, 2012.

22. Beckmann, L., Thomas, D. C., Fischer, C., & Chang-Claude, J. (2005). Haplotype sharing analysis using mantel statistics. *Human Heredity*, *59*, 67–78. doi:.10.1159/000085221

23. Wessel, J., & Schork, N. J. (2006). Generalized genomic distance-based regression methodology for multilocus association analysis. *American Journal of Human Genetics*, *79*, 792–806. doi:.10.1086/508346

24. Tzeng, J.-Y., Zhang, D., Chang, S.-M., Thomas, D. C., & Davidian, M. (2009). Gene-trait similarity regression for multimarker-based association analysis. *Biometrics*, *65*, 822–832. doi:.10.1111/j.1541-0420.2008.01176.x

25. Mukhopadhyay, I., Feingold, E., Weeks, D. E., & Thalamuthu, A. (2010). Association tests using kernel-based measures of multi-locus genotype similarity between individuals. *Genetic Epidemiology, 34,* 213–221. doi:.10.1002/gepi.20451

26. Wei, Z., Li, M., Rebbeck, T., & Li, H. (2008). U-statistics-based tests for multiple genes in genetic association studies. *Annals of Human Genetics, 72,* 821–833. doi:.10.1111/j.1469-1809.2008.00473.x

27. Goeman, J. J., van de Geer, S. A., de Kort, F., & van Houwelingen, H. C. (2004). A global test for groups of genes: Testing association with a clinical outcome. *Bioinformatics, 20,* 93–99. http://www.ncbi.nlm.nih.gov/pubmed/14693814 accessed September 24, 2012.

28. Tzeng, J.-Y., & Zhang, D. (2007). Haplotype-based association analysis via variance-components score test. *American Journal of Human Genetics, 81,* 927–938. doi:.10.1086/521558

29. Kwee, L. C., Liu, D., Lin, X., Ghosh, D., & Epstein, M. P. (2008). A powerful and flexible multilocus association test for quantitative traits. *American Journal of Human Genetics, 82,* 386–397. doi:.10.1016/j.ajhg.2007.10.010

30. Wu, M. C., Kraft, P., Epstein, M. P., Taylor, D. M., Chanock, S. J., Hunter, D. J., et al. (2010). Powerful SNP-set analysis for case-control genome-wide association studies. *American Journal of Human Genetics, 86,* 929–942. doi:.10.1016/j. ajhg.2010.05.002

31. Schaid, D. J. (2010). Genomic similarity and kernel methods I: Advancements by building on mathematical and statistical foundations. *Human Heredity, 70,* 109–131. doi:.10.1159/000312641

32. Neale, B. M., Rivas, M. A., Voight, B. F., Altshuler, D., Devlin, B., Orho-Melander, M., et al. (2011). Testing for an unusual distribution of rare variants. *PLOS Genetics, 7,* e1001322. doi:.10.1371/journal.pgen.1001322

33. Liu, J. Z., McRae, A. F., Nyholt, D. R., Medland, S. E., Wray, N. R., Brown, K. M., et al. (2010). A versatile gene-based test for genome-wide association studies. *American Journal of Human Genetics, 87,* 139–145. doi:.10.1016/j.ajhg.2010.06.009

34. Luan, Y., & Li, H. (2008). Group additive regression models for genomic data analysis. *Biostatistics, 9,* 100–113. doi:.10.1093/biostatistics/kxm015

35. Chatterjee, N., Kalaylioglu, Z., Moslehi, R., Peters, U., & Wacholder, S. (2006). Powerful multilocus tests of genetic association in the presence of gene-gene and gene-environment interactions. *American Journal of Human Genetics, 79,* 1002–1016. doi:.10.1086/509704

36. Zhao, J., Boerwinkle, E., & Xiong, M. (2005). An entropy-based statistic for genomewide association studies. *American Journal of Human Genetics, 77,* 27–40. doi:.10.1086/431243

37. Li, B., & Leal, S. M. (2008). Methods for detecting associations with rare variants for common diseases: Application to analysis of sequence data. *American Journal of Human Genetics, 83*, 311–321. doi:.10.1016/j.ajhg.2008.06.024

38. Madsen, B. E., & Browning, S. R. (2009). A groupwise association test for rare mutations using a weighted sum statistic. *PLOS Genetics, 5*, e1000384. doi:.10.1371/journal.pgen.1000384

39. Morgenthaler, S., & Thilly, W. G. (2007). A strategy to discover genes that carry multi-allelic or mono-allelic risk for common diseases: A cohort allelic sums test (CAST). *Mutation Research, 615*, 28–56. doi:.10.1016/j.mrfmmm.2006.09.003

40. Li, B., & Leal, S. M. (2009). Discovery of rare variants via sequencing: Implications for the design of complex trait association studies. *PLOS Genetics, 5*, e1000481. doi:.10.1371/journal.pgen.1000481

41. Price, A. L., Kryukov, G. V., de Bakker, P. I. W., Purcell, S. M., Staples, J., Wei, L.-J., et al. (2010). Pooled association tests for rare variants in exon-resequencing studies. *American Journal of Human Genetics, 86*, 832–838. doi:.10.1016/j.ajhg.2010.04.005

42. Han, F., & Pan, W. (2010). A data-adaptive sum test for disease association with multiple common or rare variants. *Human Heredity, 70*, 42–54. doi:.10.1159/000288704

43. Morris, A. P., & Zeggini, E. (2010). An evaluation of statistical approaches to rare variant analysis in genetic association studies. *Genetic Epidemiology, 34*, 188–193. doi:.10.1002/gepi.20450

44. Wu, M. C., Lee, S., Cai, T., Li, Y., Boehnke, M., and Lin, X. Rare-variant association testing for sequencing data with the sequence kernel association test. (2011). *American Journal of Human Genetics, 89*, 82–93. doi:.10.1016/j.ajhg.2011.05.029

45. Tzeng, J.-Y., Zhang, D., Pongpanich, M., Smith, C., McCarthy, M. I., Sale, M. M., et al. (2011). Studying gene and gene-environment effects of uncommon and common variants on continuous traits: A marker-set approach using gene-trait similarity regression. *American Journal of Human Genetics, 89*, 277–288. doi:.10.1016/j.ajhg.2011.07.007

46. Lin, D.-Y., & Tang, Z.-Z. (2011). A general framework for detecting disease associations with rare variants in sequencing studies. *American Journal of Human Genetics, 89*, 354–367. doi:.10.1016/j.ajhg.2011.07.015

47. Cai, T., Lin, X., & Carroll, R. J. (2012). Identifying genetic marker sets associated with phenotypes via an efficient adaptive score test. *Biostatistics, 13*, 776–790. doi:.10.1093/biostatistics/kxs015

48. Dai, J. Y., Kooperberg, C., Leblanc, M., & Prentice, R. L. (2012). Two-stage testing procedures with independent filtering for genome-wide gene-environment interaction. *Biometrika, 99*, 929–944. doi:.10.1093/biomet/ass044

49. Jiao, S., Hsu, L., Bézieau, S., Brenner, H., Chan, A. T., Chang-Claude, J., et al. (2013). SBERIA: Set-based gene-environment interaction test for rare and common variants in complex diseases. *Genetic Epidemiology.* doi:.10.1002/gepi.21735

50. Lin, X., Lee, S., Christiani, D. C., & Lin, X. (2013). Test for interactions between a genetic marker set and environment in generalized linear models. *Biostatistics.* doi:.10.1093/biostatistics/kxt006

51. Broderick, P., Carvajal-Carmona, L., Pittman, A. M., Webb, E., Howarth, K., Rowan, A., et al. (2007). A genome-wide association study shows that common alleles of SMAD7 influence colorectal cancer risk. *Nature Genetics, 39*, 1315–1317. doi:.10.1038/ng.2007.18

52. Tomlinson, I. P. M., Webb, E., Carvajal-Carmona, L., Broderick, P., Howarth, K., Pittman, A. M., et al. (2008). A genome-wide association study identifies colorectal cancer susceptibility loci on chromosomes 10p14 and 8q23.3. *Nature Genetics, 40*, 623–630. doi:.10.1038/ng.111

53. Tenesa, A., Farrington, S. M., Prendergast, J. G. D., Porteous, M. E., Walker, M., Haq, N., et al. (2008). Genome-wide association scan identifies a colorectal cancer susceptibility locus on 11q23 and replicates risk loci at 8q24 and 18q21. *Nature Genetics, 40*, 631–637. doi:.10.1038/ng.133

54. Peters, U., Hutter, C. M., Hsu, L., Schumacher, F. R., Conti, D. V., Carlson, C. S., et al. (2011). Meta-analysis of new genome-wide association studies of colorectal cancer risk. *Human Genetics, 131*, 217–234. doi:.10.1007/s00439-011-1055-0

55. International HapMap Consortium. The International HapMap Project. (2003). *Nature, 426*, 789–796. doi:.10.1038/nature02168

56. Huang, H., Chanda, P., Alonso, A., Bader, J. S., & Arking, D. E. (2011). Gene-based tests of association. *PLOS Genetics, 7*, e1002177. doi:.10.1371/journal.pgen.1002177

57. Lee, S., Emond, M. J., Bamshad, M. J., Barnes, K. C., Rieder, M. J., Nickerson, D. A., et al. (2012). Optimal unified approach for rare-variant association testing with application to small-sample case-control whole-exome sequencing studies. *American Journal of Human Genetics, 91*, 224–237. doi:.10.1016/j.ajhg.2012.06.007

5 A Gene-Based Approach for Testing Gene × Gene and Gene × Environment Interactions

Tao Wang

Large-scale genome-wide association studies (GWAS) offer an unprecedented opportunity to find genes underlying human diseases or traits. To date, many genome-wide significant associations have been successfully produced [1–6]. Nevertheless, the findings of GWAS explain a small proportion of the heritability for most common diseases and traits. One important possibility accounting for this phenomenon is the G × G and G × E interaction [7]. It has long been recognized that most human diseases or traits have a multifactorial etiology involving complex G × G and G × E interactions [8–13]. However, GWAS often used single marker-based tests to interrogate the association between each SNP marker and the trait of interest [14]. As the result, those variants that act synergistically with other variants and environmental factors, but have small, or even no main effects, could have been missed [15, 16]. Moreover, such analysis could also lead to inconsistent results from different populations because it fails to account for the fact that the influence of a specific genetic variant on the diseases usually relies on the genetic and environmental background.

Although the importance of the G × G and G × E interaction has been widely recognized, it is not a trivial task to account for interactions in the analysis of GWAS data. To date, few G × G and G × E interactions have been identified and replicated [17–22]. This raises possibility that current approaches used for the genome-wide G × G and G × E interaction analysis still have severe limitations. The major obstacle is multiple-testing: Searching for a large number of possible interactions requires a very stringent statistical significance level for controlling for type 1 error rate, incurring severe penalty on statistical power. To reduce multiple-testing burden, one approach is to perform a multiple-stage interaction analysis, in which a subset of SNPs is selected at the screening stage to be tested for the G ×

G and G × E interaction. The screening approach may be based on evidence of marginal effects [23], and G-E correlations in a case-control design [24]. Another approach is to leverage the information of multiple SNPs in a gene region to perform the gene-based interaction analysis. The gene-based approach has several advantages because (1) it requires multiple-testing correction based on the number of gene-sets rather than millions of SNPs in the genome; (2) it tests the accumulative effect of multiple variants; (3) it is still able to provide important biological insights as gene is the basic function unit of the genome.

One statistical challenge of using this gene-based approach relates to the large number of degrees of freedom involved in the test statistic, which could result in a severe penalty on the efficiency. For example, for genes with k SNPs, there are k possible interactions between the gene and an environmental factor, potentially leading to a test with k degrees of freedom. An ideal approach for modeling the G × G and G × E interaction should be able to make use of all the useful information from multiple SNPs and yet avoid the penalty due to a large number of degrees of freedom. One good example is based on Tukey's 1-df model for nonadditivity [25]. This procedure assumes that the SNP marker data, S, are a surrogate for an underlying quantitative biological phenotype and that the mean of this quantitative phenotype can be described by a linear function of s. As a result, a parsimonious model is built up based on the biological phenotype with only one parameter involved in modeling the interactions. This interaction parameter is estimated by searching in a prespecified region to maximize the association between the gene and the disease. Essentially, this method extracts the genotypic information of multiple SNPs of a gene region as a single component, i.e., the biological phenotype, and this compression process is made possible by correlations between the SNPs and the disease. Because of its parsimony, this procedure can greatly improve the power to detect association in the presence of interactions. However, SNPs may have no, or small, marginal effects in certain genetic models and therefore the correlations between these SNPs and the disease tend to be small and uninformative for reducing the genotype data. Nevertheless, in this case the correlation between the SNPs and the environmental exposure could be useful for dimension reduction. So, a potentially approach may be obtained by simultaneously exploiting both the disease-SNP correlations and the SNP-SNP (or the environment-SNP) correlations. This motivates us to extend the idea of the Tukey's 1-df model for developing a new gene-based method for modeling G × G and G × E interactions. We consider a regression model in which the interaction is similarly modeled by latent components. To extract an informative latent component, multiple SNP data are compressed by simultaneously exploiting the correlations between SNPs in a gene region, the environmental factor, and the disease. The latent component is defined to have a maximal correlation of SNPs in a gene region with both the disease and the SNPs in another gene region (or an environmental factor), and it is obtained by the partial least square (PLS) algorithm. We simulate case-control data to evaluate the type 1 error rate and power of this new method. We apply this method to the endometrial cancer case-control dataset.

Methods

We illustrate the methods by first considering the interaction between two gene regions, which can be readily applied for testing the G × E interaction. Consider a case-control design in which a sample of cases ($D = 1$) and controls ($D = 0$) is selected from a population. We assume two sets of SNP markers are genotyped in the two gene regions with genotype values $S_1 = (s_{12}s_{12}, ..., s_{1k_1})$ and $S_1 = (s_{21}s_{22}, ..., s_{2k_2})$. The underlying causal genetic variants may or may not be included in the two sets of observed SNPs. We consider an additive coding scheme in which the genotype value of each SNP can be 0, 1, and 2, corresponding to genotypes aa, Aa, and AA, respectively. We are interested in modeling the interaction between the two underlying variants with the aim of improving the power to detect genetic association.

To take the interaction between two underlying genetic variants into account in detecting association, a conventional method exhaustively models all pairwise interactions between the two genes by fitting a logistic regression model of the form

$$\text{logit}\{\Pr(D)\} = \beta_0 + \sum_{i=1}^{k_1} \beta_{1i} s_{1i} + \sum_{j=1}^{k_2} \beta_{2j} s_{2j} + \sum_{i=1}^{k_1} \sum_{j=1}^{k_2} \beta_{12ij} s_{1i} s_{2j} \tag{5.1}$$

We can see that a large number of parameters are used in modeling the interaction between the two genes in this way and therefore the statistic to detect association involves a large number of degrees of freedom. Moreover, although different pairs of SNPs may have quite different information to predict the pair of underlying disease variants, in this method each pair of SNPs is treated equally with one degree of freedom.

The Test Based on Tukey's Model

To avoid the severe penalty on power from having a large number of degrees of freedom, an efficient approach to model interaction should remove non-informative SNPs or weight each SNP based on its information regarding the interaction effect. One approach relies on the disease information. Given a genetic interaction model with nonnegligible marginal effects, it is reasonable to assume that a SNP marker with a larger marginal effect should have a larger interaction effect. In this way, Tukey's 1-df interaction model can be constructed so that the interaction effect is proportional to the product of two main effects; these are each described by a weighted sum of all SNPs in a region, with the weights being determined according to the SNP's correlation with the disease [25]. This model can be described by the logistic regression model

$$\text{logit}\{\Pr(D)\} = \beta_0 + \sum_{i=1}^{k_1} \beta_{1i} s_{1i} + \sum_{j=1}^{k_2} \beta_{2j} s_{2j} + \theta \sum_{i=1}^{k_1} \sum_{j=1}^{k_2} \beta_{1i} \beta_{2j} s_{1i} s_{2j} \tag{5.2}$$

Fixing the effects of the SNPs in one gene region, the interaction parameter in this model, θ, can be looked upon as a transformation parameter to remove any removable nonadditive effect of the tested gene.

Because of the parsimony of the above model, in the presence of interaction it has major advantages over both the test that does not model any interaction and the test based on model (5.1). But the gain in power of this approach depends on the assumption that a SNP's interaction effect on the disease is approximately proportional to its marginal effect on the disease. Hence, this approach may not be optimal in power when there exist no, or only small, marginal effects.

The Test Based on Principal Component Analysis

Similar to Tukey's 1-df model, we can assume the disease status depends on two underlying genetic phenotypes, u_1 and u_2, in two gene regions through a linear model with interaction as given by equation 5.1.

$$logit\{Pr(D)\} = \theta_0 + \theta_1 u_1 + \theta_2 u_2 + \theta_{12} u_1 u_2 \tag{5.3}$$

In contrast to the approach based on model (5.2), one can define the underlying genetic phenotype by relying solely on the genotype data themselves. One common approach is to use principal component analysis (PCA). Consider p SNPs in a gene region evaluated on each member of a sample of n persons so that the multiple genotype values make up a $p \times n$ matrix $\mathbf{S}$. The singular value decomposition of $\mathbf{S}$ is $\mathbf{S}=\mathbf{UDV}^{\mathrm{T}}$, where $\mathbf{D}$ is a diagonal matrix containing the singular values, and the elements of the column vector $\mathbf{U}$ are the principal components $U^1, U^2, ..., U^m$ (in which $m \leq p$ is the rank of $\mathbf{S}$). One may simply use the first principal components as u_1 and u_2, but the method can be easily extended to include more components. However, arbitrarily choosing top principal components may not be optimal as it may miss important information. Another approach is to define the underlying genetic phenotype by weighting principal components by their corresponding eigenvalues. For the sake of simplicity, here we use the first principal component and a parsimonious interaction model between the two genes can be described by

$$logit\{\Pr(D)\} = \beta_0 + \sum_{i=1}^{k_1} \beta_{1i}s_{1i} + \sum_{j=1}^{k_2} \beta_{2j}s_{2j} + \beta_{12}U_1^1 U_2^1 \tag{5.4}$$

This model also has only one degree of freedom in modeling the interaction between two genes. Because this principal component analysis depends solely on the genotype data $\mathbf{S}$, it can also be used to reduce the dimension for modeling main effects without introducing bias in the subsequent tests [26]. Here, we focus on modeling the interaction between two genes. The rationale for the test based on principal component analysis is that, since the first principal components are able to capture most of the variation in the genotype data in the two regions, it is reasonable to assume that the genotypic variation introduced by the genetic interaction effects is likely to be included in the first components.

Because the PCA approach characterizes the pairwise linkage disequilibrium (LD) pattern of genotype data but does not rely on correlations of SNPs with the disease, it may be able to detect interactions that the approach based on the marginal effect of each SNP could otherwise miss. However, the logistic regression based on the first few components, such as model (5.4), might not always work well. Because the first principal components are constructed based only on the genotypic correlation structure, the efficiency of this approach is sensitive to the LD pattern of the SNP markers.

We can illustrate this problem with the extreme LD pattern shown in figure 5.1. The genotype values for each SNP in a gene region are displayed by a row and the genotype values for an individual are given by a column in figure 5.1 (top). The three genotype values are coded white (0), gray (1), and black (2). It can be seen that the subset of SNPs labeled "A" have higher genotype values in the first and second blocks of subjects and have lower values in the third and fourth blocks of subjects, whereas the subset labeled "B" have higher values in the first and third blocks and lower values in the second and fourth blocks. PCA is able to uncover the structure of this genotype data. The set of SNPs (A) that shows the largest variation contributes most to the first principal component (figure 5.1, bottom). However, in this example the set of SNPs (A) is in fact not informative for detecting association. From this fictitious example, we can see that the first few principal components may explain the genotype data in the genes rather than the disease, and so nothing guarantees that the first principal components are consistently informative relative to the disease in the various possible LD patterns.

The Test Based on PLS

It is unlikely that the preceding genotype dimension reduction methods, either the approach based on the marginal effects of single SNPs or the approach based only on the local pairwise LD structure, are uniformly powerful in different situations. One approach may be highly powerful in certain situations, but it may also have no power at all in other situations. Unfortunately, it is unclear which one should be used in a real dataset, because the power of these methods depends on the unknown genetic model.

It is therefore useful to have an approach to model the G × G and G × E interaction effect that can generally perform well in different situations. The idea is that, rather than directly exploit the LD structure of SNP markers in a gene region as PCA does, we reduce multiple SNP genotypes into one component by using information from both the marginal effects of the SNPs and the joint effects on the disease of all pairs of SNPs in the two genes.

The marginal effect of a SNP can be characterized by the correlation between its genotype value and the disease. Several methods have successfully used this correlation to reduce the burden of multiple testing for

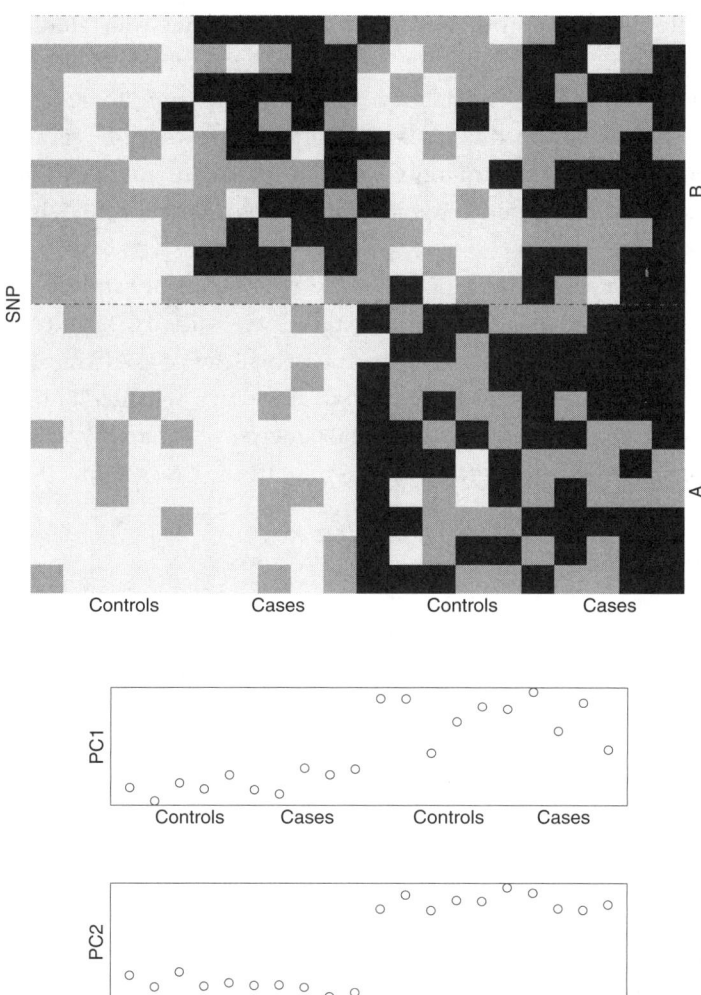

Figure 5.1

The genotype type values in a fictitious gene region. A: the genotype values—the three genotype values are coded white (0), gray (1), and black (2); B: distribution of the first two principal components of the fictitious data.

testing interactions. For example, in the scenario of genome-wide studies, Marchini et al. consider a two-stage approach in which a subset of selected SNPs is further modeled for interaction [23]. The criterion to remove uninformative SNPs relies on this disease-genotype correlation. In certain genetic models with no or small marginal effects, it is also useful to explore the joint effect of SNPs in order to reduce the genotype data for modeling $G \times G$ and $G \times E$ interactions.

To investigate the joint effect of SNPs for a case-control sample, we assume there are two independent underlying disease variants, u_1 and u_2, one in each gene region. Let A_1 and a_1 be the two alleles of a SNP marker that is in LD with u_1 and A_2 and a_2 be the two alleles of a SNP marker that is in LD with u_2. Let $f_{i_1 j_2 = P(D|i_1 j_2)}$ be the marginal penetrance of haplotype $i_1 j_2$, where i and j can each be one of the alleles A or a. In the case group,

$$P_{a_1 a_2}^D = P(a_1, a_2 \mid D) = \frac{P(D \mid a_1 a_2) P(a_1 a_2)}{P(D)} = \frac{P_{a_1} P_{a_2} f_{a_1 a_2}}{P_D}$$

$$P_{A_1 a_2}^D = \frac{P_{A_1} P_{a_2} f_{A_1 a_2}}{P_D}$$

$$P_{a_1 A_2}^D = \frac{P_{a_1} P_{A_2} f_{a_1 A_2}}{P_D}$$

$$P_{A_1 A_2}^D = \frac{P_{A_1} P_{A_2} f_{A_1 A_2}}{P_D}$$

LD between two causal loci in cases is expected and is given by

$$\Delta_{12}^D = P_{a_1 a_2}^D P_{A_1 A_2}^D - P_{A_1 a_2}^D P_{d_1 D_2}^D = \frac{P_{a_1} P_{A_1} P_{a_2} P_{A_2} (f_{a_1 a_2} f_{A_1 A_2} - f_{a_1 A_2} f_{A_1 a_2})}{P_D^2}$$

From this equation, we can see that the joint effect of two causal variants will lead to LD between alleles of the variants in cases as long as $f_{a_1 a_2} f_{A_1 A_2} - f_{a_1 A_2} f_{A_1 a_2} \neq 0$. Furthermore, the correlation between two markers that are in LD with the two causal variants can also be observed, and its expectation is given by

$$\Delta_{s_1 s_2}^D = \frac{\Delta_1 \Delta_2}{P_{a_1} P_{A_1} P_{a_2} P_{A_2}} \Delta_{12}^D \qquad (5.5)$$

where Δ_1 and Δ_2 are the LD coefficients between the two markers and the two variants, respectively [27]. Similarly, the joint effect of two disease variants introduces LD between alleles of the markers and the two causal variants in the controls, but this has an opposite sign, and is given by

$$\Delta_{12}^{\bar{D}} = \frac{P_{a_1}P_{A_1}P_{a_2}P_{A_2}[(1-f_{a_1a_2})(1-f_{A_1A_2})-(1-f_{a_1A_2})(1-f_{A_1a_2})]}{(1-P_D)^2}$$

$$\Delta_{s_1s_2}^{\bar{D}} = \frac{\Delta_1\Delta_2}{P_{a_1}P_{A_1}P_{a_2}P_{A_2}}\Delta_{12}^{\bar{D}} \tag{5.6}$$

From equations (5.5) and (5.6), we can see that, even when the marginal effects of the causal variants are small, for example in a crossover model ($f_{a_1a_2}$ and $f_{A_1A_2} > f_{a_1A_2}$ and $f_{A_1a_2}$), the LD between SNPs in two genes can be large. In a case-control study, especially for rarer diseases, cases are usually oversampled, and the LD in controls tends to be closer to the population LD and smaller. Hence, although cases and controls have opposite signs of LD, we can still expect LD between SNP markers in two gene regions for the whole sample of cases and controls whenever two causal variants have a joint effect on the disease. This suggests that the LD between two genes can be a useful alternative source of information to reduce the genotype data for modeling interactions when marginal effects are small. We note that, because this genotype reduction does not make use of disease status information, under the null hypothesis the reduced genotype data are independent with disease status and therefore the subsequent test is still valid [24].

The joint effect of two SNPs may be directly characterized by the LD coefficient. However, the LD coefficients cannot be estimated directly, because the haplotype phase of two SNPs in two genes is unknown. Although the haplotype phase may be inferred, e.g., by the expectation-maximization (EM) algorithm with the assumption that the two SNPs are in Hardy-Weinberg equilibrium (HWE), the HWE is likely to be distorted in samples in the region with causal variants. To avoid such an assumption, the composite LD measures are useful alternatives. The composite LD is estimated simply as half the usual sample covariance of the genotype values and the composite LD correlation is estimated by a sample correlation. Similarly, the SNP-environment correlation can be a useful source of information to reduce the genotype data for modeling gene-based G × E interaction.

To incorporate information from the SNP-disease correlation and the correlations between SNPs of two gene regions in modeling interactions, one may use a PLS procedure. Let $\mathbf{S}_1$ be an $N \times k_1$ genotypic matrix of N cases and controls and k_1 SNPs in the gene region g_1 and let $\mathbf{D}$ denote the $N \times 1$ vector indicating disease status. We are interested in detecting association between the gene g_1 and disease status. Also, let $\mathbf{S}_2$ be the genotypic matrix for another gene region g_2 (this could instead be a matrix of the observed values of environmental factors). In PCA, the principal component may be defined sequentially by:

$$\arg \max_{\|v\|=1} var(\mathbf{S}_1 v)$$

where $\mathbf{v}$ is the weight and the first component U_1^1 is the linear combination that has the maximum variance. As discussed before, PCA maximizes the variance of the linear combination of multiple genotype values and may not necessarily yield a component that is informative for modeling interaction because it ignores the information from $\mathbf{D}$ and $\mathbf{S}_2$. For this reason, a different criterion is necessary for extracting a component that is informative for modeling interactions. How well this genetic component represents u_1 depends on: (1) the magnitude of its correlation with the disease, if there exists a marginal effect of the underlying variant u_1; and (2) the magnitude of its correlation with the SNPs in the other region g_2, if there exists a joint effect of the two causal variants. In other words, a useful component has to be defined on the basis of information from the correlations between the SNPs in regions g_1 and g_2, and the marginal correlations between SNP and the disease.

Based on these criteria, we consider another linear combination, $\mathbf{S}_1 \mathbf{w}$, to represent u_1. Letting $\mathbf{Z} = (\mathbf{D}, \mathbf{S}_2)$, which combines the vector $\mathbf{D}$ and the matrix $\mathbf{S}_2$, the weight vector $\mathbf{w}$ can be naturally defined by

$$\arg \max_{\|\mathbf{w}\|=1, \|\mathbf{c}\|=1} cov^2(\mathbf{S}_1\mathbf{w}, \mathbf{Z}\mathbf{c})$$

The weights $\mathbf{w}$ and $\mathbf{c}$ can be found by a PLS algorithm [28, 29]. PLS is now provided by commonly used statistical packages, such as SAS and R. Although PLS can sequentially find more orthogonal components, in our application we use only the first component U_1^{1pls} to estimate the underlying interaction component u_1. The PLS component is also a linear combination of genotype values of SNPs in a local region, but we can see that the PLS and PCA yield different weights for the SNPs, based on the different criteria of PCA and PLS, namely maximizing $var(\mathbf{S}_1 v)$ and $cov^2(\mathbf{S}_1\mathbf{w}, \mathbf{Z}\mathbf{c})$, respectively. It is useful to understand the PLS component given by different genetic models. If u_1 has no marginal effect, the component is obtained by giving more weight to SNPs that have stronger absolute composite LD correlations with the SNPs in g_2. If a larger marginal effect exists, the component is defined by giving more weights to SNPs having larger absolute correlations with the disease. So the PLS component is more likely to capture information from multiple SNPs than a PCA component for the aim of modeling interactions. Based on this informative component, we can apply conventional logistic regression by replacing the principal components with PLS components in model (5.4) to efficiently test the interaction effect and the overall association. In summary, our procedure for detecting association of a gene region with multiple SNPs genotyped is as follows: (1) Standardize genotypes and disease phenotypes to have zero mean and unit norm; (2) Find U_1^{1pls}; (3) Detect association by a likelihood ratio test based on a regression model of the form (4) using PLS components.

Simulation Study

We evaluate the performance of the proposed method and several other approaches by simulations. We simulate two candidate gene regions, g_1 and g_2, with seven SNPs in each. In each region one of the simulated SNPs is assumed to be the causal variant and the others are the SNP markers. We assume the causal variant and other SNP markers are in LD. The haplotypes of correlated SNP markers and the causal variant are simulated based on a multivariate normal distribution with pairwise correlations ρ_{ij}, where i and j are the position indices. The absolute value of ρ_{ij} is set to be $0.95^{|j-i|}$ and the sign of ρ_{ij} is randomly assigned. We assume that the disease variant is located at the first position. Each allele of a haplotype is generated by dichotomizing the marginal normal distribution and the cutoff is determined by the allele frequency. The allele frequency of a SNP is randomly sampled from a uniform distribution between 0.1 and 0.5. This simulation yields correlations between SNPs roughly in a range of 0.1 to 0.75.

The disease status of each subject is simulated based on a logistic-regression model,

$$P(D=1) = \frac{exp[\theta_0 + \theta_1 I_{u1} + \theta_2 I_{u2} + \theta_{12} I_{u1} I_{u2}]}{1 + exp[\theta_0 + \theta_1 I_{u1} + \theta_2 I_{u2} + \theta_{12} I_{u1u2}]}$$

where I_{u1} and I_{u2} are indicator variables for the disease alleles at the two causal loci. We fix the intercept parameter, θ_0, which corresponds to a disease probability for a group of subjects who do not carry any disease alleles, to be 0.018. For each replicate, we generate a homogenous study population and then randomly sample a balanced dataset with 400 cases and 400 controls.

To evaluate the type 1 error rate of the different approaches to detect association between g_1 and the disease, we consider two situations: (1) neither u_1 nor u_2 are associated with the disease $\theta_1 = \theta_2 = \theta_{12} = 0$ and (2) only u_2 is related to the disease $\theta_2 = 2, \theta_1 = \theta_{12} = 0$. For each scenario we use 10,000 replicates to evaluate the type 1 error rate.

To compare the power of the different methods, we consider three different genetic models. In a multiplicative model, the joint effect of two factors, u_1 and u_2, is simulated to be the product of their main effects ($\theta_{12} = 0$, θ_1 and $\theta_2 > 0$). In a purely epistatic model, the main effects of u_1 and u_2 are simulated to be 0 $\theta_1 = \theta_2 = 0$ and the joint effect of the two factors is solely determined by the interaction effect ($\theta_{12} > 0$). In a crossover model, we assume u_1 has opposite effects depending on the genotype u_2. We fix $\theta_1 = -0.5$ and let $\theta_{12} > 0$. We consider two further scenarios: (1) the causal variants u_1 and u_2 are not genotyped, i.e., the first SNP is removed from the analysis; (2) the causal variants are genotyped, in which case the last SNP is removed from the analysis, so that in each case there are consistently six SNPs in the analysis.

Type 1 Error Rate

Simulations show that the proposed test has good control of the 1 percent and 5 percent error rates when the number of markers in the two genes is six (table 5.1). However, we find evidence that the statistic based on a full regression model is anticonservative. Because the number of parameters involved in this method is large, the poor control of the type 1 error rate of the statistic is likely the result of small sample violations of asymptotic

Table 5.1
The empirical type 1 error rates of various methods to detect association of gene 1 at the significance levels 0.05 and 0.01

	$\alpha = 0.05$		$\alpha = 0.01$	
	$\theta_2 = 0$	$\theta_2 = 2$	$\theta_2 = 0$	$\theta_2 = 2$
Main	0.051	0.052	0.011	0.011
Full	0.094	0.102	0.023	0.027
PCA	0.052	0.048	0.013	0.009
Tukey's 1-df	0.049	0.047	0.010	0.010
PLS	0.051	0.049	0.011	0.011

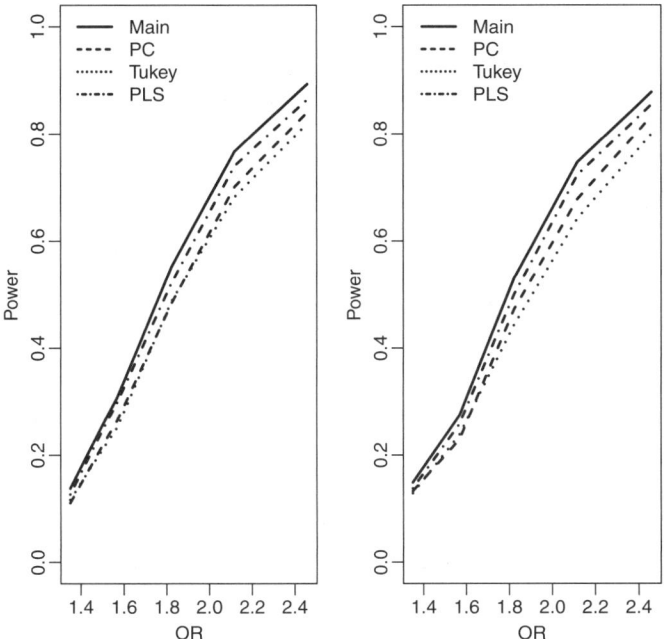

Figure 5.2
The empirical power to detect association of the disease with gene 1 under a multiplicative model. The causal variants are not genotyped. (Left) The allele frequency of U2 is 0.1; (Right) The allele frequency of U2 is 0.2.

theory. Therefore, we increased the sample size and then found the type 1 error of the statistic tended to be close to the nominal level.

Power

Figures 5.2–5.4 show the empirical power of different methods for testing association of the disease and the underlying causal variant u_2 at a significance level of $\alpha = 0.05$ when the causal variant is not genotyped. Because the test based on a full logistic regression does not have good control of the type 1 error rate and is generally much less powerful than the other approaches, the results of this approach are not shown.

When the genetic model is multiplicative (figure 5.2), the standard main effect logistic regression (Main) is most powerful, as expected. However, the proposed approach (PLS), the score test based on the first principal component (PC) and the score test based on Tukey's 1-degree of freedom (df)

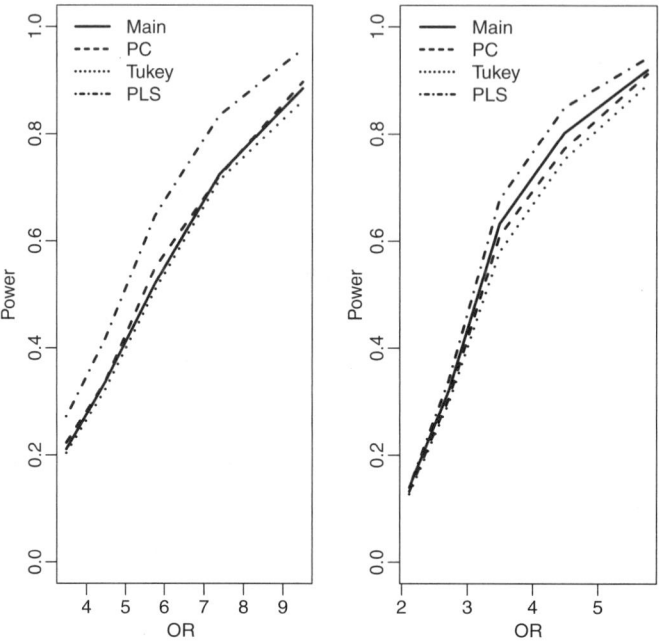

Figure 5.3
The empirical power to detect association of the disease with gene 1 under an epistatic model. The causal variants are not genotyped. (1) The allele frequency of U2 is 0.1; (2) The allele frequency of U2 is 0.2.

model (Tukey) have only slightly less power than the main effect logistic regression model, because only one degree of freedom is unnecessarily used for modeling interaction. In contrast, the logistic regression model with full pairwise interactions between two SNPs in two genes significantly loses power because of the penalty on the power due to the large number of degrees of freedom for modeling interactions (data not shown). We can also see that, because we use a likelihood ratio test rather than a score test, there is a minor difference in power among PLS, PC, and Tukey's 1-df method. When the underlying genetic model is purely epistatic (figure 5.3), the new PLS approach clearly has the best performance among all approaches. The full logistic regression again has the worst performance. Interestingly, we can see that the PC and Tukey's 1-df approaches have power very similar to that of the main effect logistic regression model. The power of a test statistic that models interactions depends on a trade-off between the number of

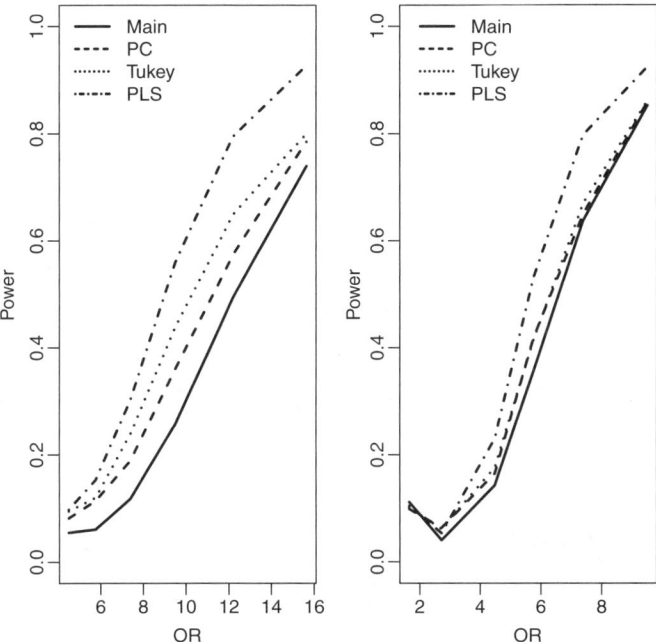

Figure 5.4
The empirical power to detect association of the disease with gene 1 under a crossover model. The causal variants are not genotyped. (1) The allele frequency of U2 is 0.2.

degrees of freedom and the interaction effects of the SNP markers. As the correlations between the underlying disease variants and the genotyped SNPs are moderate in this simulation, the interaction effects of the SNPs are small and so the PC and Tukey's 1-df approaches do not outperform the main effect logistic regression. However, the PLS approach outperforms the other approaches by taking advantage of information from both the disease and the genotypic correlation structure. We can see that the gain in power of PLS depends on the allele frequency of u_2. When the allele u_2 is rarer, the gain in power is larger, which was also observed by Chatterjee et al. in examining the power of Tukey's 1-df approach [25]. Figure 5.4 shows the empirical power of the various methods as a function of θ_{12} when the underlying genetic model is a crossover model ($\theta_1 = -0.5$). We can see that, because of the modest marginal effects, the main effect logistic regression has much lower power than the PLS, PC and Tukey's 1-df methods, among

Table 5.2
The p-values of different approach in testing AKT1 and AKT2 associated with endometrial cancer by taking their interactions with BMI into account

	AKT1	AKT2
Main	0.083	0.026
PC	0.097	0.102
Tukey	0.103	0.022
PLS	0.042	0.002

which the PLS has the best power performance. Similarly, the power advantage of the PLS is observed in the simulation scenarios that the underlying disease variable are genotyped.

Application: A Study of AKT2, MAPK3, and Risk of Endometrial Cancer

Cancer of the endometrium is the most common gynecologic malignancy and accounts for 6 percent of all cancers in women. Estimated new cases and deaths from endometrial cancer in the United States in 2007 are 39,080 and 7,400, respectively. Estrogen replacement therapy (ERT) in menopause and obesity are the principal risk factors for endometrial cancer [30, 31]. Although only a small proportion of women using ERT or with obesity will develop endometrial cancer, little is known about genetic susceptibility to the disease.

To illustrate the method proposed here, we have applied it to the data from a genetic study of endometria cancer. This study has investigated several candidate genes, including AKT1 (MIM164730) and AKT2 (MIM164731), in 246 endometrial cancer cases and 688 controls. AKT1 and AKT2 are members of the AKT family, which is relevant to tumorigenesis because of its effect on anti-apoptosis and cell proliferation [32].

In this study, six and seven tag-SNPs are genotyped for AKT1 and AKT2, respectively. The frequencies of the minor SNP alleles range from 0.04 to 0.45. The different approaches are applied to detect association of AKT1 and AKT2 with endometrial cancer taking their interactions with BMI into account. Table 5.2 presents the p-values of the different multi-locus analysis approaches. The results from our approach suggest that AKT1 is associated with endometrial cancer susceptibility at the significance level 0.05. However, the main-effect test and other approaches fail to detect association between AKT1 and endometrial cancer. The nonsignificant results of other approaches suggest that the marginal effects of SNPs in AKT1 are small. All tests, except PC, show significant association between AKT2 and cancer, for which the new PLS method tends to show slightly stronger evidence of association than the other approaches, which is consistent with what we observed in the simulation studies.

Dimension Reduction on the Interaction Terms

In the above PLS approach, the dimension reduction is performed on the genotype values of SNPs to obtain the underlying genetic phenotype for modeling the interaction. Another approach of using the PLS is to perform the dimension reduction directly on the interaction terms between SNPs in two gene regions. Let S_{12} be an $N \times k_1 k_2$ interaction matrix of N cases and controls with k_1 and k_2 SNPs in the gene region g_1 and g_2. The PLS components can be defined by

$$\arg \max_{\|\mathbf{w}\|} \text{cov}^2(\mathbf{S}_{12}\mathbf{w}, \mathbf{D})$$

The interaction between two genes can then be tested by examining the association between the top PLS components and disease. It is useful to note a connection between this PLS method and the test of the interaction effects under a variance component model. Under the variance component model, the interaction effects in model (1), β_{12ij}s, are assumed to be random effects with a common variance. The score test of the variance component may be simply written as

$$T = \mathbf{D}^\mathrm{T}\mathbf{S}_{12}\mathbf{S}_{12}^\mathrm{T}\mathbf{D}.$$

This score test is closely related to the sum of squared statistics (SSU) [33], which can be looked as a special case of the gene-environment set association test (GESAT) [34]. Since the first component of the PLS is $\mathbf{S}_{12}\mathbf{S}_{12}^\mathrm{T}\mathbf{D}$, the score test of the variance component of the interaction effects can be viewed as a test for correlation between the first PLS component and disease.

Discussion

Modeling G × G and G × E interactions can not only improve the power for detecting association of genetic markers with diseases but can also be useful for understanding the underlying mechanisms of diseases. However, the prevailing analysis strategy for genetic association studies still follows a locus-by-locus strategy. As an alternative, the gene based interaction approach is appealing as it can reduce the multiple-testing burden and improve power in the presence of multiple disease variants. Although a conventional logistic regression can be readily applied, a large number of parameters have to be used in modeling all possible pairs of SNPs, potentially leading to an unsuccessful test.

To improve power, it is necessary to reduce the dimension of interaction terms in the logistic regression. The PLS approach can be performed either on the genotype matrix of a gene or on the interaction matrix of two genes (or one gene and an environmental factor). The PLS approach based on the genotype matrix is similar to the procedure based on Tukey's 1-df model in that they both reduce the dimension of multiple SNPs into a single latent variable to capture the useful information. As such this approach can have only one degree of freedom involved in modeling the interaction and hence least cost in power when the true genetic model is purely multiplicative. When the underlying genetic model is an interaction model in which the effect of a causal variant relies on the genotypes of another variant, this approach potentially has a major power advantage over a procedure that does not take interactions into account, because large effects exist for each genotype group of the other variant, although the marginal effect of the causal variant can be small. The PLS approach is different from the procedure based on Tukey's 1-df model in that in this procedure the dimension reduction of multiple SNPs relies on both the marginal effects of SNPs and the LD between the two genes in the sample. Because of the use of both sources of information, it is likely to have a more robust performance. In an extreme example, in which a causal variant has opposite effects according to different genotypes of the other variants, so that the marginal effects of both variants do not exist, the proposed approach can work well because the LD patterns are used to obtain an informative latent component. However, the interaction effect of the causal variant may not be captured by Tukey's 1-df model, so that the procedure based on Tukey's 1-df model may not have good power. Another advantage of the PLS method is that the interaction effect and the overall effect (both the main and interaction effects) of a gene can be directly tested by applying a standard logistic regression, but how to test the interaction effect is not obvious in the procedure based on Tukey's 1-df model. When the PLS procedure is performed on the interaction matrix between two genes (or a gene and an environmental factor), this approach is closely related to SSU [33] and GESAT [34], two other gene-based approaches. It is useful to compare the advantages and disadvantages of the PLS approaches performed on the genotype matrix and the interaction matrix in various scenarios and to develop a new method that can have the advantages of both approaches.

An alternative approach to reduce SNPs in modeling G × G and G × E interactions follows a two-stage procedure in which SNPs that meet some threshold in a test at the first analysis phase are subsequently followed up for modeling interactions. The selection of SNPs may be simply based on a test to examine the correlation between SNPs and the phenotype, which relies on the existence of nonnegligible marginal effects [23, 35]. Millstein et al. have described another alternative method: First, select SNPs based on a test for the significance of correlations in pairwise SNPs [24]. However, the correlation test is usually less efficient than the marginal effect test when nonnegligible marginal effects exist. The usefulness of these two approaches may vary for different underlying genetic models. Moreover, it is not clear how to choose the threshold for selection in this two-stage procedure. The PLS method and the test based on Tukey's 1-df model take a "soft" shrinkage approach in which SNPs are given different weights, so that the threshold does not need to be specified in advance.

The major assumption of our approach is that the two genes are not in LD in the population. When this assumption is not valid, the new test may not achieve maximal power because the LD introduced by the disease is confounded with the background LD, although the test based on this approach can still keep good control of the type 1 error rate. In this case, it may be useful to estimate shrinkage means of the genotype values to remove the "background" LD [36]. Another limitation of our approach, as with all other approaches to model interactions, is that the efficiency to detect interactions decreases rapidly with decreasing LD between the marker and the causal variants. Without a satisfactory coverage of SNPs, the gain in power of the proposed method tends to be small even in a purely epistatic model, although it does not lead to a significant loss of power. To successfully detect association of genes that have no, or small, marginal effects, denser markers are desirable than in a design for detecting marginal effects.

The gene-based method can handle any number of SNPs in a gene, but it may not be wise to blindly apply this approach to a large number of SNPs because accumulated noise from unrelated SNPs could mask any real association. Before applying this method, it is useful to first examine the LD pattern of the gene. When SNPs tend to be in different LD blocks, one component may not be enough to capture all the information. One may consider applying this approach to each block or using more than one component in modeling interactions.

It is straightforward to use the above method to test gene-based G × G and G × E interactions for a continuous trait. However, the SNP-SNP (or SNP-environment) correlations may not be informative in dimension reduction of genotype values for a continuous trait, because the interaction between two genes (or a gene and an environmental factor) does not introduce correlations when individuals are randomly sampled from the population. Nevertheless, the marginal effects of SNPs can still be available. Essentially, the PLS method is to assign different weights to SNPs of a gene. For a continuous trait, the phenotype variances may be different between genotype groups of a SNP in the presence of interaction effect of the SNP [37, 38]. This can be a useful source of information to weight SNPs when the marginal effects are small.

Acknowledgments

This research was supported by National Institute of Health grants 1R21HL118637and R21HG006150.

References

1. Anonymous. (2007). Genome-wide association study of 14,000 cases of seven common diseases and 3,000 shared controls. *Nature, 447*(7145), 661–678.

2. Saxena, R., Voight, B. F., Lyssenko, V., Burtt, N. P., de Bakker, P. I., Chen, H., et al. (2007). Genome-wide association analysis identifies loci for type 2 diabetes and triglyceride levels. *Science, 316*(5829), 1331–1336.

3. Scott, L. J., Mohlke, K. L., Bonnycastle, L. L., Willer, C. J., Li, Y., Duren, W. L., Erdos, M. R., et al. (2007). A genome-wide association study of type 2 diabetes in Finns detects multiple susceptibility variants. *Science, 316*(5829), 1341–1345.

4. Sladek, R., Rocheleau, G., Rung, J., Dina, C., Shen, L., Serre, D., et al. (2007). A genome-wide association study identifies novel risk loci for type 2 diabetes. *Nature, 445*(7130), 881–885.

5. Steinthorsdottir, V., Thorleifsson, G., Reynisdottir, I., Benediktsson, R., Jonsdottir, T., Walters, G. B., et al. (2007). A variant in CDKAL1 influences insulin response and risk of type 2 diabetes. *Nature Genetics, 39*(6), 770–775.

6. Zeggini, E., Weedon, M. N., Lindgren, C. M., Frayling, T. M., Elliott, K. S., Lango, H., et al. (2007). Replication of genome-wide association signals in UK samples reveals risk loci for type 2 diabetes. *Science, 316*(5829), 1336–1341.

7. Manolio, T. A., Collins, F. S., Cox, N. J., Goldstein, D. B., Hindorff, L. A., Hunter, D. J., et al. (2009). Finding the missing heritability of complex diseases. *Nature, 461*(7265), 747–753.

8. Aston, C. E., Ralph, D. A., Lalo, D. P., Manjeshwar, S., Gramling, B. A., DeFreese, D. C., et al. (2005). Oligogenic combinations associated with breast cancer risk in women under 53 years of age. *Human Genetics, 116*(3), 208–221.

9. Barlassina, C., Lanzani, C., Manunta, P., & Bianchi, G. (2002). Genetics of essential hypertension: From families to genes. *Journal of the American Society of Nephrology, 13*(Suppl 3), S155–S164.

10. Hsueh, W. C., Cole, S. A., Shuldiner, A. R., Beamer, B. A., Blangero, J., Hixson, J. E., et al. (2001). Interactions between variants in the beta3-adrenergic receptor and peroxisome proliferator-activated receptor-gamma2 genes and obesity. *Diabetes Care, 24*(4), 672–677.

11. Kim, J. H., Sen, S., Avery, C. S., Simpson, E., Chandler, P., Nishina, P. M., et al. (2001). Genetic analysis of a new mouse model for non-insulin-dependent diabetes. *Genomics, 74*(3), 273–286.

12. Kuida, S., & Beier, D. R. (2000). Genetic localization of interacting modifiers affecting severity in a murine model of polycystic kidney disease. *Genome Research, 10*(1), 49–54.

13. Naber, C. K., Husing, J., Wolfhard, U., Erbel, R., & Siffert, W. (2000). Interaction of the ACE D allele and the GNB3 825T allele in myocardial infarction. *Hypertension, 36*(6), 986–989.

14. Manolio, T. A., Brooks, L. D., & Collins, F. S. (2008). A HapMap harvest of insights into the genetics of common disease. *Journal of Clinical Investigation, 118*(5), 1590–1605.

15. Culverhouse, R., Suarez, B. K., Lin, J., & Reich, T. (2002). A perspective on epistasis: limits of models displaying no main effect. *American Journal of Human Genetics, 70*(2), 461–471.

16. Thomas, D. (2010). Gene-environment-wide association studies: Emerging approaches. *Nature Reviews. Genetics*, *11*(4), 259–272.

17. Dimitriou, M., & Dedoussis, G. Z. (2012). Gene-diet interactions in cardiovascular disease. *Current Nutrition Reports*, *1*(3), 153–160.

18. Joseph, P. G., Pare, G., & Anand, S. S. (2013). Exploring gene-environment relationships in cardiovascular disease. *Canadian Journal of Cardiology*, *29*(1), 37–45.

19. Qi, L. (2012). Gene-diet interactions in complex disease: Current findings and relevance for public health. *Current Nutrition Reports*, *1*(4), 222–227.

20. van Vliet-Ostaptchouk, J. V., Snieder, H., & Lagou, V. (2012). Gene-lifestyle interactions in obesity. *Current Nutrition Reports*, *1*, 184–196.

21. Franks, P. W., Pearson, E., & Florez, J. C. (2013). Gene-environment and gene-treatment interactions in type 2 diabetes: Progress, pitfalls, and prospects. *Diabetes Care*, *36*(5), 1413–1421.

22. Garcia-Rios, A., Perez-Martinez, P., Delgado-Lista, J., Lopez-Miranda, J., & Perez-Jimenez, F. (2012). Nutrigenetics of the lipoprotein metabolism. *Molecular Nutrition & Food Research*, *56*(1), 171–183.

23. Marchini, J., Donnelly, P., & Cardon, L. R. (2005). Genome-wide strategies for detecting multiple loci that influence complex diseases. *Nature Genetics*, *37*(4), 413–417.

24. Millstein, J., Conti, D. V., Gilliland, F. D., & Gauderman, W. J. (2006). A testing framework for identifying susceptibility genes in the presence of epistasis. *American Journal of Human Genetics*, *78*(1), 15–27.

25. Chatterjee, N., Kalaylioglu, Z., Moslehi, R., Peters, U., & Wacholder, S. (2006). Powerful multilocus tests of genetic association in the presence of gene-gene and gene-environment interactions. *American Journal of Human Genetics*, *79*(6), 1002–1016.

26. Gauderman, W. J., Murcray, C., Gilliland, F., & Conti, D. V. (2007). Testing association between disease and multiple SNPs in a candidate gene. *Genetic Epidemiology*, *31*(5), 383–395.

27. Zhao, J., Jin, L., & Xiong, M. (2006). Test for interaction between two unlinked loci. *American Journal of Human Genetics*, *79*(5), 831–845.

28. Holland, I. S. (1988). On the structure of partial least-squares regression. *Communications in Statistics. Simulation and Computation*, *17*, 581–607.

29. Holland, I. S. (1990). Partial least-squares regression and statistical-models. *Scandinavian Journal of Statistics*, *17*, 97–114.

30. Kaaks, R., Lukanova, A., & Kurzer, M. S. (2002). Obesity, endogenous hormones, and endometrial cancer risk: A synthetic review. *Cancer Epidemiology, Biomarkers & Prevention, 11*(12), 1531–1543.

31. Pike, M. C., & Ross, R. K. (2000). Progestins and menopause: Epidemiological studies of risks of endometrial and breast cancer. *Steroids, 65*(10–11), 659–664.

32. Altomare, D. A., & Testa, J. R. (2005). Perturbations of the AKT signaling pathway in human cancer. *Oncogene, 24*(50), 7455–7464.

33. Pan, W. (2010). Statistical tests of genetic association in the presence of gene-gene and gene-environment interactions. *Human Heredity, 69*(2), 131–142.

34. Lin, X., Lee, S., Christiani, D. C., & Lin, X. (2013). Test for interactions between a genetic marker set and environment in generalized linear models. *Biostatistics, 14*(4), 667–681.

35. Evans, D. M., Marchini, J., Morris, A. P., & Cardon, L. R. (2006). Two-stage two-locus models in genome-wide association. *PLOS Genetics, 2*(9), e157.

36. Wang, T., Zhu, X., & Elston, R. C. (2007). Improving power in contrasting linkage-disequilibrium patterns between cases and controls. *American Journal of Human Genetics, 80*(5), 911–920.

37. Struchalin, M. V., Dehghan, A., Witteman, J. C., van Duijn, C., & Aulchenko, Y. S. (2010). Variance heterogeneity analysis for detection of potentially interacting genetic loci: method and its limitations. *BMC Genetics, 11*, 92.

38. Pare, G., Cook, N. R., Ridker, P. M., & Chasman, D. I. (2010). On the use of variance per genotype as a tool to identify quantitative trait interaction effects: A report from the Women's Genome Health Study. *PLOS Genetics, 6*(6), e1000981.

6 An Overview of the RELIEF Algorithm and Advancements

Alexandre Todorov

The aim of the RELIEF algorithm and derivatives is to identify features (e.g., genes, environmental factors) that are *relevant* to the trait of interest (e.g., risk increasing or decreasing), starting from a set of that may include thousands of irrelevant features (e.g., SNP genotypes from a whole-genome chip). The RELIEF algorithm and derivatives have been applied in numerous settings, ranging from document identification [1] to distinguishing between objects, persons, and animals in pictures, modeling electricity price spikes [2], and, recently, in gene-expression profiling [3, 4] and studies of epistatic (gene-gene) interactions [5]. To the best of our knowledge, its application to the study of gene-environment interaction studies has been limited. This may be due in part to the relatively low publicity the algorithm received in genetic epidemiological journals until recently; but also to the fact that RELIEF is less geared toward detecting interaction effects than toward detecting genetic effects (interactive or not) *in the presence* of complex interactions and other confounding factors. We will argue that there is currently a great deal of value in the later.

The aim of this chapter is to introduce RELIEF and some of its numerous variants. We begin with a broad summary of the aims and scope of RELIEF, the problems it addresses, and some of the outstanding issues. We then describe the classic RELIEF algorithm and the more recent variants. A comparison of RELIEF and logistic regression is made afterward. We thus focus on a single technique—a more general discussion of machine learning applications in genetics can be found in Cordell [6].

Scope of RELIEF

The first thing to emphasize is that RELIEF is not designed for the *detailed* modeling of gene-environment interaction effects. RELIEF will not provide, say, an estimate of interaction effect sizes. In fact, the philosophy of the approach is to identify relevant features *despite* potential confounding factors such as complex interactions (e.g., gene-environment, epistatic) or nonlinear relationships between features and outcome. Such factors may have contributed significantly to the failure of other approaches to identify gene-environment interactions in genome-wide studies.

Kira and Rendell (1992a, 1992b), who formulated the original algorithm [7, 8], intended the approach to be used as a *screener* to identify a subset of features that may not be the smallest and may still include some irrelevant and some redundant features, but that is small enough to use with more refined approaches in a detailed analysis—they mention ID3, the precursor of the popular C5.0 algorithm. These more refined approaches would simply be unable to handle very large numbers of features, most of which are irrelevant. In that spirit, Moore and colleagues (2013), for example, offer RELIEF as a preliminary filter prior to analysis in their MDR package for the analysis of gene-gene and gene-environment interaction [9]. Recent work suggests this approach is more effective than penalized logistic regression to detect gene-gene interactions [10, 11]. RELIEF is also used as one possible filter in the Encore epistasis network analysis toolkit [12].

Advantages of RELIEF

RELIEF is a completely nonparametric approach. It makes no assumption regarding the relationship between features and the outcome. Most important, it retains the ability to identify relevant features with minimal marginal (main) effects [13]. A classic example is interaction models that have no associated main effects, which would be identified by RELIEF but not in the usual "massively univariate" analysis [5]. This does not mean that it is totally "model proof." For example, Bins and Draper note that non-monotonic relationships (e.g., age and 12-month drug-use liability, with an inverted-U shape) would reduce the power of RELIEF [14]; but the same would be true of a logistic based analysis that did not take the nonlinearity into account.

Redundant Features

An advantage of RELIEF is that it is not hampered by multicollinearity among the candidate features, in contrast to other approaches. For example, the *lasso* is a popular type of forward-selection regression, where the number of predictors introduced in the model is controlled by adding a penalty proportional to the sum of the absolute value of the regression coefficients [15, 16]. It is known that the *lasso* will select, almost randomly, between two highly correlated predictors, whereas RELIEF will return both. Returning redundant features is a benefit to some extent, as it may point to meaningful clusters of correlated phenotypes. Alternatively, Yang and Li proposed, roughly stated, a combination of RELIEF and principal component analysis to address this issue [17].

Difficulties with RELIEF

RELIEF produces relevance estimates for each feature that are typically constrained to be between −1 and 1 (larger positive indicating higher relevance). The distribution of these statistics is complex and explicit forms will exists only for trivially simple cases. Consequently, despite being cited as one of the most successful algorithms in its class, there has been little work on the *statistical* properties of RELIEF. If the algorithm is clearly successful in certain situations (e.g., image processing), it is still unclear exactly how it will perform when applied to the fuzzier data sets encountered in psychiatric genetic epidemiology and G × E studies. The fact that RELIEF and its derivatives were typically developed by experts in computer science may have also contributed to the lower prioritization of the study of the statistical properties of RELIEF, the interest being more on the (nontrivial!) aspects of algorithm development and implementation, and convergence issues.

In particular, the question remains how to attach significance levels to the feature relevance estimates. Kira and Rendell proposed a simple rule for significance testing, based on Chebyshev's inequality, which proves unsatisfactory in practice. The most viable alternative may simply be permutation testing [18]. There, however, computation costs become quite significant. Sun et al., for example, note that their version of *RELIEF* can work with millions of irrelevant features [19] (not a far cry in this era of whole-genome sequencing), but the companion MATLAB code carries the warning, "prepare to be patient."

The power of the RELIEF algorithm is yet to be determined. Kira and Rendell, the designers of the RELIEF algorithm, remarked that the algorithm is valid only when *"the relevance level is huge for relevant features and small for irrelevant features."* Huge effects are rare in genetic epidemiology.

But it is comforting to know that promising results from the application of RELIEF to genetic data sets accumulate [12, 20].

Software Implementation

RELIEF is available through the MDR package [13]. Robnik-Šikonja and colleagues developed the R package CoreLearn, which also has a stand-alone C++ implementation and has the advantage of providing other filtering algorithms to allow cross-method comparisons [21]. Sun et al. [19] provide the iterative relief algorithm using Matlab. The examples presented here were produced using a local C++ library freely available from me. It uses the open-source libraries RandomLib for number generation [22] and DLIB for optimization [23].

Classic RELIEF

Original RELIEF

The RELIEF algorithm (figure 6.1) as formulated in Kira and Rendell [7, 8] is used rarely now. Nevertheless, it is still worthwhile to present their

0. Initialization

    0.1   Select J observations to serve as targets and hold their indices in list T. Options include:
- Sample with or without replacement
- sample all observations rather than some
- Sample from one class only (e.g., all targets are affected)

        Initialize the F feature relevance weights to 0, $\omega_f = 0$.

1. Main iteration loop

    For each target t in the list T

    1.1   Calculate the distance beween t and all other observations

    1.2   Find h = nearest hit, nearest observations from the same class as t
            Find m = nearest miss, nearest observation from a different class than t

    1.3   Increment the relevance estimates. For each feature f,

$$\omega_f \leftarrow \omega_f + \frac{1}{J}\left(d\left(x_{tf}, x_{mf}\right) - d\left(x_{tf}, x_{hf}\right)\right)$$

2. End

    2.1   Rank-order the features in terms of ω. The most relevant features (predicted) have the largest $\omega_f > 0$. The least relevant features have $\omega_f \leq 0$.

    2.2   (optional) Repeat analysis with a different list of targets to verify stability of solution

Figure 6.1
Pseudocode of classic RELIEF (Kira and Rendell, 1992).

original formulation, as much of the underlying principles and the original structure have been retained [24]. The premise of RELIEF is, sensibly, that discordant pairs (one person affected and the other unaffected) should be more different on a relevant feature (e.g., a risk increasing mutation) than a concordant pair (either both affected, or both unaffected).

The inputs to RELIEF are the class assignments (**Y**), and the $N \times F$ feature matrix (**X**) where x_{if} denotes feature f for individual i. The output is a vector of F estimates of feature relevance, ω. The relevant features (predicted) have the largest $\omega_f > 0$; the least relevant features have $\omega_f \leq 0$. The features can be numerical or nominal, numerical features being scaled to $(0,1)$ prior to the analysis (subtracting the minimum value and dividing by the range). The core element of RELIEF is the difference function. It is typically:

$$d(x,y) = \begin{cases} |x-y| & \text{feature is ordinal or continuous} \\ x \neq y & \text{nominal features} \end{cases},$$

although any valid distance metric could be used—Kira and Rendell, for example, use the Euclidian distance.

Now let the indices t, h, and m refer, respectively to: "target" (an observation chosen at random); "nearest hit," the observation from the same class as the target (e.g., both affected) and closest to it; and "nearest miss," the closest observation to t, from a class other than the target's class. Precisely, we should be using $h(t)$ and $m(t)$ to emphasize the fact we are considering hits and misses nearest a specific target. This is not done to alleviate notation. For each target and each feature, define the difference score $Z_{tf} = d(x_{tf}, x_{mf}) - d(x_{tf}, x_{hf})$, which ranges from -1 to $+1$. It is $+1$ when target and miss differ, and target and hit are similar (increasing the relevance estimate of the feature); and -1 when target and miss are similar, and target and hit are different (decreasing the relevance estimate of the feature). Suppose now that we have selected J targets at random. A natural estimate of feature relevance is then:

$$\omega_f = \frac{1}{J} \sum_{t \in T} \left(d(x_{tf}, x_{mf}) - d(x_{tf}, x_{hf}) \right) = \frac{1}{J} \sum_{t \in T} Z_{tf} \qquad (6.1)$$

This equation is in a nutshell the RELIEF algorithm as proposed by Kira and Rendell. ω_f may theoretically range from -1 (evidence against relevance) to $+1$ (most relevant). These values are unlikely in practice: a $+1$ would imply that all concordant pairs share the same feature and that all discordant pairs have different feature values.

The simplicity of the approach is clear. The only parameters that the user must specify in the original RELIEF algorithm are: (1) J, the number of targets; and (2) the difference function to use. The consensus is that a large number of targets is preferable, and several authors actually recommend dispensing with random target selection and using each observation once as a target. It is not yet clear that this is the optimal strategy, since different observations may have different information content. As regards the choice of distance function, this is not the most crucial element; for example, Euclidian (as in Kira and Rendell) and Manhattan distances yield comparable results.

Interpretation of the Relevance Weights

For nominal and binary features, Robnik-Šikonja and Kononenko (2003) interpret ω_F as the estimate of the difference between (1) the probability that the target and the nearest miss differ on a relevant attribute (high if the attribute is relevant) and (2) the probability that the target and nearest miss differ on that attribute (low, if the attribute is relevant) [25]. In this case, $Z_{tf} = d\left(x_{tf}, x_{mf}\right) - d\left(x_{tf}, x_{hf}\right)$ in equation 6.1 will equal -1, 0, or $+1$. Tests for the comparison of multinomial distributions could then be used to test the significance of the relevance estimate, since under the null hypothesis (the feature is not relevant to the disease), Z_{tf} takes on the values -1, 0, $+1$ with probabilities $p(1-p)$, $1-2\,p(1-p)$ and $p(1-p)$ where p is the item frequency.

This interpretation of relevance weights is not wholly satisfactory, since it does not apply to continuous features. This complicates not only their interpretation but would also seem to invalidate one key principle of RELIEF, that features can be ranked in terms of relevance using the ω_f 's only. Bachrach-Gilad et al. [26] and Sun et al. [17, 27] provide an interesting analysis of RELIEF showing why in fact such ranking is of interest, by pointing out the relationship between the relevance weights that RELIEF calculates and the "margin" in a 1–NN classifier. Assume for a moment that the class of the target is unknown and let the "margin" for target t:

$$\rho_t = \sum_f Z_{tf} = \sum_f \left(d\left(x_{tf}, x_{mf}\right) - d\left(x_{tf}, x_{hf}\right)\right).$$

In a 1–NN classifier, the target t would assigned (correctly) to the class of h when $\rho_t > 0$; and to the class of m when $\rho_t < 0$. This observation led to a number of improvements that will be discussed later.

The Difficulty of Initially Selecting Nearest Neighbors

Although we clearly see the value of RELIEF, we spent some time on the issue of distance, as one could not expect the method to work when huge amounts of irrelevant features are included in the analysis. When a high number of irrelevant features are used in distance computations, the distance between two observations is, for all practical purposes, independent of their affection status. The performance of RELIEF can significantly degrade in this situation, where "hits" and "misses" are essentially selected at random. To illustrate this point, suppose that a disease is determined by a mutant allele and an environmental exposure. Denote by P_{uv} the probability that an affected individual has exposure $x_e = u$ and mutant carrier status $x_a = v$ (u, v = 0,1). Let Q_{uv} denote the corresponding probabilities in unaffected individuals, $P_{\cdot v} = P_{0v} + P_{1v}$, and $Q_{\cdot v} = Q_{0v} + Q_{1v}$. Table 6.1 summarizes the probabilities of the relevance weights for the genetic factor, under random selection of hits and misses. Note that, when hits and misses are selected at random, only the marginal probabilities come into play—an interaction effect, no matter how strong, does not contribute to the evidence for A.

As an example, consider the extreme situation where carrying the mutant and being exposed are necessary and sufficient for the disease and there are no sporadic cases. The prevalence of the disease under this model is $Q_{11} = p_A p_E$, the product of the frequencies for the mutant and the exposure. In figure 6.2, for each combination of p_A and p_E that yield a prevalence of 1 percent, we plotted the expected $d(x_{tf}, x_{mf})$ versus the expected $d(x_{tf}, x_{hf})$, first, for randomly selected hits and misses given an affected target (red) or an unaffected target (blue); and, second, when only nearest hits and nearest misses are used in the analysis (green, for the unaffected target case only). This highlights three key points. First, it is possible to have *negative* evidence if we chose hits (or misses) at random from the subsets of individuals

Table 6.1

Distribution of relevance weights if select hits and misses at random

Target	$P(\omega_a = 1)$	$P(\omega_a = -1)$
Affected	$P_{\cdot 0}^2 \cdot Q_{\cdot 1} + P_{\cdot 1}^2 \cdot Q_{\cdot 0}$	$2P_{\cdot 0}P_{\cdot 1}$
Unaffected	$Q_{\cdot 0}^2 \cdot P_{\cdot 1} + Q_{\cdot 1}^2 \cdot P_{\cdot 0}$	$2Q_{\cdot 0}Q_{\cdot 1}$

*The case $\omega_a = 1$ is obtained by subtraction, $1 - P(\omega_a = +1) - P(\omega_a = -1)$

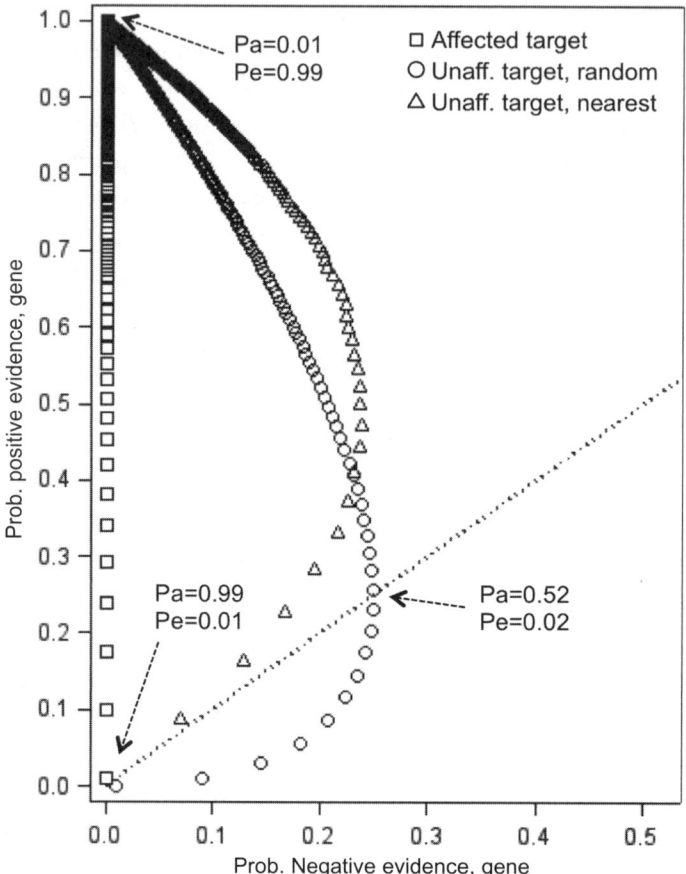

Figure 6.2

An example where using randomly selected hits and misses instead of nearest hits and misses could lead to the wrong inference (model descripted in the text). The core of RELIEF is the difference between the probabilities that the target is different from a miss (positive evidence) and the probability that it is different from a hit (negative evidence). If hits and misses are chosen at random, the magnitude and direction of this difference depends on the choice of target (circles: Unaffected targets; squares: Affected targets); and may be negative for some combinations of environmental exposure (*Pe*) and genetic diathesis (*Pa*) frequencies. For unaffected targets, selecting the nearest hits and the nearest miss greatly improves the situation (triangles), the difference being positive in all instances (though sometimes small).

from the same (or different) class as the target, even though the mutant is clearly relevant in the etiology of the trait. Second, the class of the target clearly matters. In this instance, affected targets (red) are far more informative than unaffected ones. Indeed, for $p_A > 0.5$ and $p_E < 0.02$, unaffected targets would actually tend to provide evidence against the relevance of the mutant. This issue, the choice of targets for analysis, has not received proper attention in the literature. Third, using nearest neighbors clearly increases the value of unaffected targets across the whole range of possible parameter values.

Derivatives of RELIEF

Observation Weighting

There are two primary directions along which RELIEF evolved: (1) the manner in which neighbors are weighted in the analyses; and (2) the shift from a single pass over the data (as in the original RELIEF) to iterative versions of the algorithm (figure 6.3). As the examples in the next section will show, there are strong advantages for each approach.

The "core" of RELIEF is, roughly stated, the averaging of the differences Z_{tf}. The more recent versions of the algorithm replace the unweighted sums used in Kira and Rendell by weighted sums,

$$Z_{tf} = \sum_{i \in M} \mu_i \cdot d(x_{tf}, x_{if}) - \sum_{j \in H} \eta_j \cdot d(x_{tf}, x_{jf}).$$

The weights μ_m and η_h are normalized to sum to 1, either within group or over the entire sample. In the original RELIEF, only one nearest neighbor was used per group, so that $\mu_m = 1$ and $\eta_h = 1$ if m and h are the nearest misses and hits; and zero otherwise. Results were found unstable and susceptible to outliers. In response, Kononenko (1997) suggested using K neighbors, so that, e.g., $\mu_m = 1/K$ if m is one the K nearest misses, and 0 otherwise (similarly for η_m) [28].

Yang et al. noted that even with more neighbors, the results may still fluctuate, if only because of the stochastic aspects of sorting (ties in distances) [29]. They suggest averaging relevance estimates over permutations of the data. A more efficient approach was suggested in Draper et al. [30] and Greene et al. [5], where one selects all neighbors within a prespecified distance r from the target. The difference between the two is that Greene et al. assign equal weights to each neighbor (e.g., the weight for near misses

0. Initialization

    0.1   Set feature *relevance*, $\theta = 0$

              Initialize feature *weights* φ_f

- $\varphi_f = 1/F$ (each feature is equally weighted), or
- $\varphi_f = 0$ for some features, = some positive value for features thought *a priori* to be relevant.

    0.2   Select J observations to serve as targets and hold their indices in list T. Options include:

- Sample with or without replacement
- sample all observations rather than some
- Sample from one class only (e.g., all targets are affected)

1. Main iteration loop

    1. For each target t in the list T:

    1.1   Define H = hits, set of observations that have the same class as t

              Define M = misses, set of observations that have a different class than t

    1.2   Calculate distances $d_{it} = \sum_f \varphi_f \cdot d\left(x_{if}, x_{tf}\right)$

    1.3   For each member of H and M, calculate observation weights (μ for misses and η for hits).

    1.4   For each feature f and target t, calculate the margin:

$$Z_{tf} = \sum_{j \in M} \mu_j \cdot d\left(x_{tf}, x_{jf}\right) - \sum_{i \in H} \eta_j \cdot d\left(x_{tf}, x_{if}\right)$$

    1.5

2. Update / End

    2.1   Single-iteration *RELIEF*: Calculate the relevance weights,

$$\omega_f = \frac{1}{N} \sum_t Z_{tf}$$

              and stop. The most relevant features (predicted) have the larges $\omega_f > 0$.

    2.2   Iterative RELIEF: Use the estimated ω to update the distance weights φ and return to step 1 if change is meaningful.

Figure 6.3

Pseudocode—*RELIEF*-type algorithms.

are $\mu_m = 1/K$ when $\delta_{tm} < r$ and zero otherwise; K is the number of near misses within a distance r of the target). In contrast, Draper et al. place more weight on the closest neighbors and less on those near the margin of that sphere, setting $\mu_m \propto \max(0, 1 - \delta_{tm}^2 / r^2)$. In both instances, the number of hits and misses will vary from target to target. This implies that the user should also specify a minimum number of targets, K_{min}, should r be too small and too few neighbors selected. The general recommendation for K_{min} is 10 to 20. Greene et al. show how to select the tuning parameter r, but the process is not trivial.

In contrast, Sun et al. [19] forgo the specification of the number of neighbors altogether and use a sigmoid weighing function over the entire set of hits (and misses):

$$\mu_{tm} = \frac{e^{-\delta_{tm}/\sigma}}{\sum\limits_{i \in M} e^{-\delta_{ti}/\sigma}} \qquad \eta_{th} = \frac{e^{-\delta_{th}/\sigma}}{\sum\limits_{j \in H} e^{-\delta_{tj}/\sigma}}.$$

The tuning parameter σ must be chosen carefully. If set too large, for example, each observation would be given nearly equal weights, which would be highly detrimental.

An alternative approach to weight formulation is quite interesting, weighting both near and distant neighbors (similar in spirit to combining affected sibpairs and extreme discordant sibpairs in linkage analysis). Unlike the approaches mentioned above that discount distant observations, or drops them altogether, Greene et al. [31] give significant weight to such distant observations, but in a manner opposite to that for near neighbors: where, for example sharing one factor for *near* hits increases relevance, the same sharing for far observations should decrease it. The weight for *near misses* are then set to $\mu_m = 1$ when $\delta_{tm} < r$ and weight for *far misses* are $\mu_m = -1$ when $\delta_{tm} > r$, where r is a prespecified threshold. The change in sign implies, for example, that far misses will contribute positively to the relevance scores of attribute on which they differ, and this is the source of the information gain in their approach. Stokes and Visweswaran [24] take the method one step further, applying a sigmoid weighting function that ranges from +1 (very close to target) to–1 (far from target):

$$w_{ti} = \frac{\theta_{ti}}{Q} \qquad \theta_{ti} = \frac{2}{1 + \exp\left(-\dfrac{4}{\psi} \cdot (\tau - \delta_{ti})\right)} \qquad Q = \sum\limits_{i=1,N} |\theta_{ti}|.$$

They suggest setting the tuning parameter τ to the average distance between pairs of observations; and ψ to the standard deviation of these distances.

Parameter Updates

In iterative versions of RELIEF, a parameter update is essentially a change in the manner in which the individual features are weighted during distance calculations. That is, in

$$d_{it} = \sum\limits_f \varphi_f \cdot d\left(x_{if}, x_{tf}\right),$$

the distance weights are initially set to $\varphi_f = 1$, since one sees no reason a priori to favor one attribute over another. One pass with RELIEF is then done on the data as described before. In classical RELIEF analyses, the analysis stops there—there is no update. In the iterative version of RELIEF, Moore and White [32] suggest setting feature weights to 0 (dropping the item) when the relevance score falls below a prespecified threshold (expressed as percentage of the remaining features). This has the major advantage of removing a number of irrelevant features from the distance calculations, and thus allowing a better estimate of relevant distance. Note that it is possible to drop an item from the distance calculation but still estimate a relevance score for it (and thus, perhaps reintroduce it at a later time). However, this will frequently cause oscillatory behavior and poor convergence, as an item with low or negative evidence is unlikely to suddenly find relevance.

Sun and Li [27] and Sun et al. [19] approached the problem from a different perspective. In the former, the new feature weights at iteration $i + 1$ are the solution to

$$\varphi^{(i+1)} = \arg\max \varphi^T \cdot \sum_{t=1}^{N} \mathbf{z}_t = \varphi^T \cdot \mathbf{v} \text{ s.t. } \|\varphi\|_2 = 1 \text{ and } \varphi \geq 0,$$

where the differences $\mathbf{Z}_t = [Z_{t1}, ..., Z_{tf}]$ are calculated using the current $\varphi^{(i)}$, weighted distances and observation weights as described earlier. As demonstrated in Sun and Li, there is a simple closed-form solution: $\varphi^{(i+1)} = \mathbf{v}^+ / \|\mathbf{v}^+\|_2$ where $v_f^+ = \max(0, v_f)$. In effect, the weight of a feature is set to zero if the current estimate of relevance is negative. This closely mirrors the suggestion in Moore and White (2007), the difference being that, in the next iteration, Moore and White use equal weights for the remaining features in the distance calculations, whereas Sun and Li use differentially weighted features. The approach proposed in Sun et al. (2010) is even more powerful. The update they propose is the solution to

$$\varphi^{(i+1)} = \arg\min \sum_t \log\left(1 + e^{-\varphi^T \mathbf{z}_t}\right) + \lambda_1 \|\varphi\|_1 + \lambda_2 \|\varphi\|_2 \text{ s.t. } \varphi \geq 0$$

We have taken some liberties with the penalty term, turning it into the popular "elastic net" [33]. In their initial formulation, Sun et al. (2010) begin with an L1 (*lasso*) penalty, but then reparametrize the problem in terms of φ^2, in effect turning it into an L2 ("ridge") penalty. They also propose a gradient-descent algorithm for the estimation procedure. We have found that this is less efficient than to use constrained limited memory

BFGS with an efficient line search [34], especially since the search can be limited to the box by $0 \leq \varphi \leq 1$. This seems justified because what matters is not the value of a feature weight, per se, but rather their weights relative to one another. However, as Sun et al. also note, the optimization is not the time consuming part of the analyses. Most of the computing time is spent in calculating the distances.

Other Directions

This discussion of RELIEF focused on binary outcomes. Kononenko and colleagues show how RELIEF can be modified to handle multiple classes without pooling classes, as pooling could reduce power by creating a heterogeneous group [28]. RELIEF is not well suited for continuous *outcomes*, and approaches to handle such outcomes [35] are ad hoc at best. Berretta et al [36] recently adapted RELIEF to handle survival data, which are commonly found in psychiatric epidemiology. Finally, we should note that McKinney and colleagues recently proposed a novel iterative approach that combines RELIEF with measures of entropy [37, 38]. It is mentioned as a promising method but not discussed here, since it is a significant departure from the traditional RELIEF approaches.

Illustration

Outline of the Simulations

As illustration, we chose a model in which a common genetic diathesis (frequency 40 percent) is expressed only when an environmental factor (E) exceeds a threshold T. We assumed that E was distributed standard normal, that $T = 1$. The probability of disease is set at 0.10 for both noncarriers and carriers of the mutant when $E < T$. For $E > T$, the penetrance remains at 0.10 for noncarriers, but increases to 0.20 for carriers. We are thus considering a situation where there is very low power to detect an environmental or genetic effect via logistic regression: The odds-ratios for the main effects are ~1.0 for the environmental factor and 1.1 for the genetic diathesis, and a modest 1.2 for the interaction.

To make the situation more challenging, we assume random sampling and a sample size of 1000. In the analyses, we add 100, 1,000, and 5,000 noise markers (unrelated to the disorder), with minor allele frequencies selected at random to be between 0.20 and 0.50. We also add 20 noise environmental factors (unrelated to affection status), which we assumed to be independently distributed standard normal.

We compared the performance of logistic regression, single-iteration RELIEF (with 10 or 20 neighbors), single-iteration RELIEF with observations exponentially weighted (setting $\sigma = 0.1$); and iterative RELIEF as in Sun et al. [17]. As noted, the iterative versions of RELIEF have one or more tuning parameters, which when, properly fitted, result in an increase in power. However, this makes them less tractable in simulation studies. We thus provide two examples, a smaller one in which the tuning parameters are optimized, and a larger set where we fix $\lambda_1 = 0$ and $\lambda_2 = 1$.

Is RELIEF Competitive?

For each of 200 simulated samples, we ran a logistic regression with the correct environmental effect and the correct genetic marker, and interaction term. Significance was assessed using Wald statistics. We further stacked the odds in favor of the logistic regression approach by selecting, for each replicate, the smaller of the p-values associated with the main genetic and the interaction effect. On the same 200 samples, we ran the classical RELIEF once with 10 and a second time with 20 equally weighted nearest-hits and misses. The success of RELIEF was assessed by the relative rank assigned to the genetic diathesis, when compared to those for 500 noise markers and 20 noise environmental covariates. A rank of 1 (best) means that the estimate was the largest among these 520 and a rank of 0 (worst) means it was the smallest.

RELIEF clearly outperforms logistic regression, even though neither method does well in this difficult situation. Even with a very a loose significance threshold of 0.001, logistic regression succeeds in only two instances to identify either the gene or a gene-environment interaction (figure 6.4). RELIEF does much better, with the gene ranked 13 percent of the time in the top 5 percent of features (~top 6 out of 520).

Advantage of Using Weighted Distance

We then assessed the benefit of using exponentially scaled distances (with $\sigma = 0.01$) as in Sun et al. (2010), to using unscaled distances with a fixed number of nearest hits and nearest misses as before (figure 6.5). The use of scaled distances results in significant improvements and provides better results 91 percent of the time compared to using 10 nearest hits and misses; and in 86

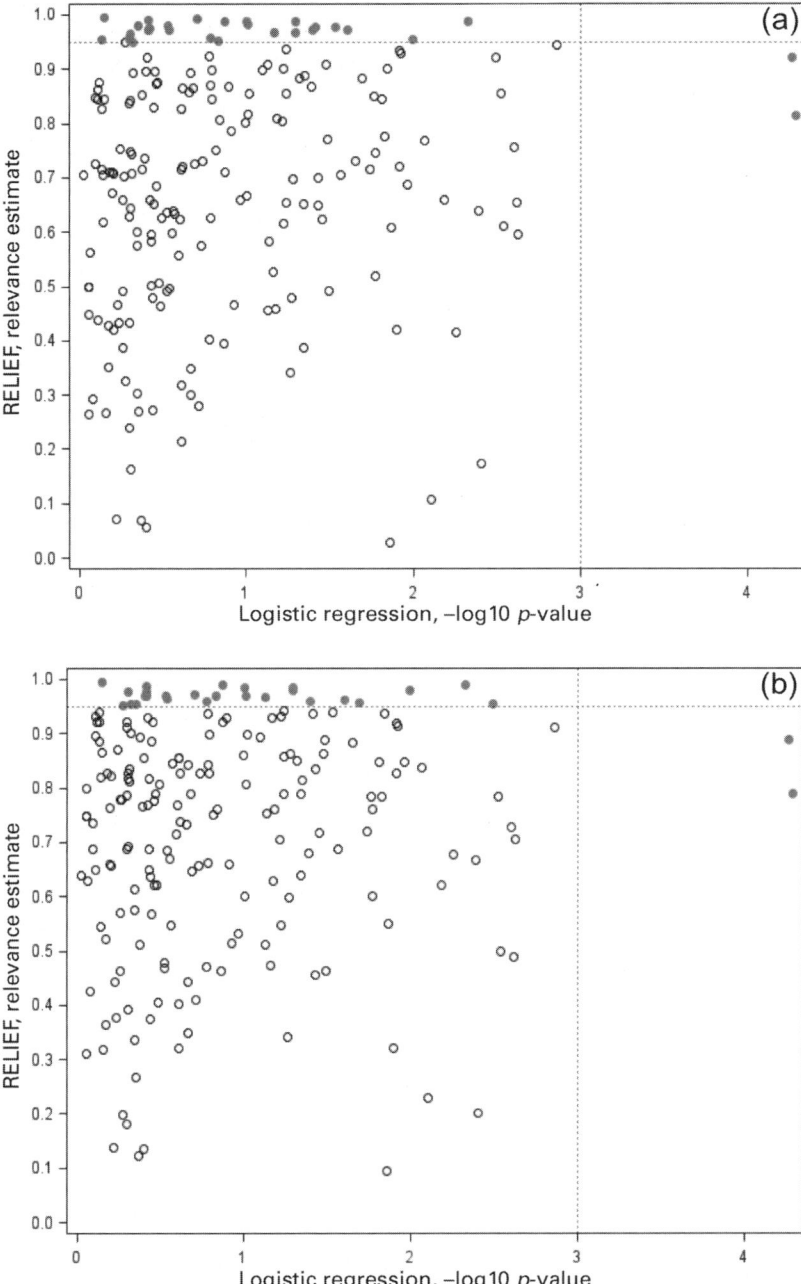

Figure 6.4

Comparison of single-iteration RELIEF and logistic regression results, for (a) $n = 10$ or (b) $n = 20$ nearest neighbors (simulation model detailed in text). The x-axis is negative log10 of the smaller of the p-values for the genetic and G x E effects from a logistic regression. The y-axis is the rank of the relevance estimates assigned to the disease locus by a single iteration of RELIEF. There are 20 noise environmental factors and 500 noise markers. The vertical line is for $p = 0.001$, which would allow a substantial number of false positives. The horizontal line is the 95 percent cutoff, allowing ~6 false-positives. Power is low, as expected, but RELIEF succeed in ~13 percent of cases compared to < 1 percent for logistic regression.

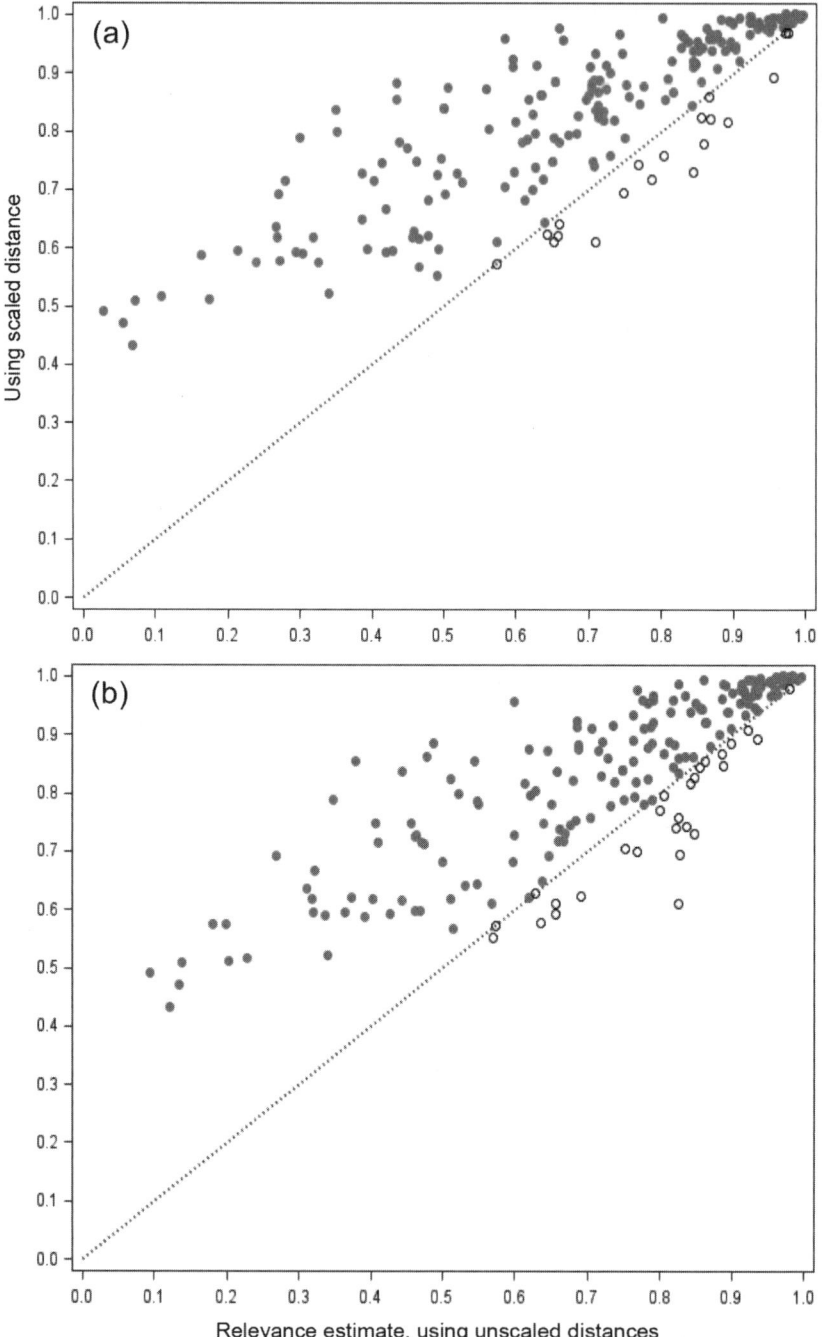

Figure 6.5

Comparison of the relevance estimates for the disease locus, with a single iteration of RE-LIEF using exponentially scaled distances ($\sigma = 0.01$) or unscaled distances with (a) $n = 10$ or (b) $n = 20$ nearest neighbors. Feature relevance estimates are scaled from 0 (irrelevant) to 1 (most relevant) among a set of features with 20 noise environmental continuous predictors and 500 noise markers. The use of scaled distances results in significant improvements and provides better results 91 percent of the time compared to $n = 10$ and 86 percent of the time compared to $n = 20$. The simulation model is described in the text.

percent of the time compared to using 20. This is particularly significant, given that the use of scaled distances adds marginally to computing time.

Advantage of Using Iterative RELIEF

Next, we considered whether iterative RELIEF, with parameter tuning as in Sun et al. (2010), can improve on a single-iteration RELIEF with 20 nearest misses and hits. We selected results from 200 iterations where the single-iteration RELIEF had failed to place the genetic factor in the top 5 percent of features (figure 6.5). The gain is clear: Iterative RELIEF (with parameter tuning) was able to identify the gene as relevant in 13 percent of these difficult instances. This added benefit, however, comes at a heavy computing cost.

Performance in the Presence of Many Noise Factors

The performance of RELIEF degraded significantly in the presence of large numbers of noise markers (we tested with 5,000; figure 6.6). However, we do note that a single iteration of iterative-RELIEF with an L2-penalty only and no parameter tuning ($\lambda_1 = 0$, λ_2 fixed at 1) has a performance comparable to a univariate screen with logistic regression. Interestingly, the analysis shows that, on the same sample, the performance of RELIEF and logistic regression can be quite divergent, suggesting that these two approaches pick up on different aspects of the data. Power would be increased to ~30 percent taking results when either RELIEF or logistic regression places the genetic factor in the top 5 percent, at the cost of ~250 false positives.

Discussion and Future Directions

If anything, the precipitous drop in the cost of genome-wide chips due to the pressure from improving sequencing technologies indicates that the need for filtering algorithms such as RELIEF will remain significant. Prior work with RELIEF and derivatives suggests that the approach has potential, and that recent improvements to the algorithm have enhanced its value. There are five areas for further research that could be pursued.

First, iterative RELIEF is essentially a "backward selection" scheme, where irrelevant features are gradually eliminated. It would be of high interest to assess performance in "forward" selection mode, where the initial distances are calculated with respect to, e.g., environmental effects that are known to be relevant. Second, performance degrades with increasing number of

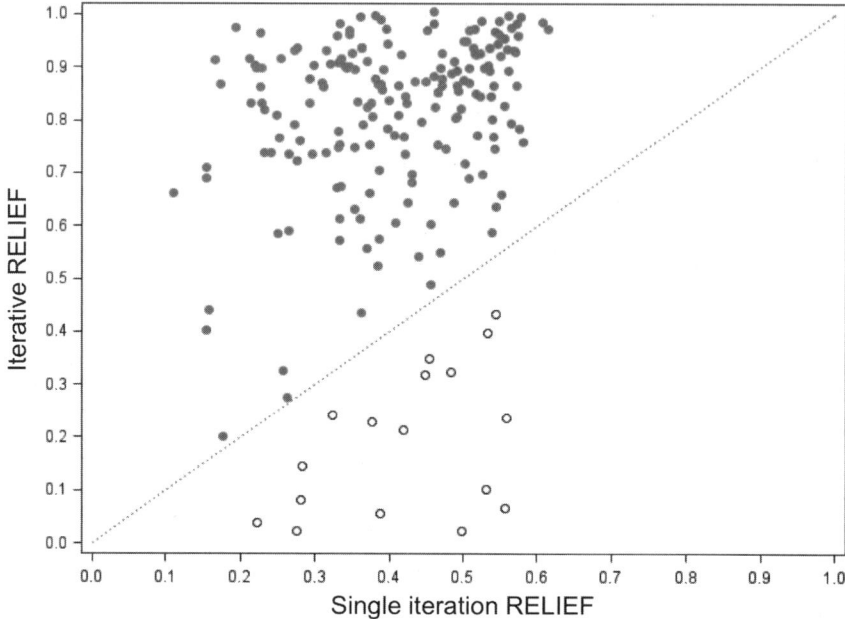

Figure 6.6

Dramatic example of how fine-tuned iterative RELIEF can improve on single-itera-tion RELIEF (20 nearest misses and hits). Improvements in the relevance rank for the genetic factor are found in 91 percent of cases. Where single iteration relief fails in all instances to place the genetic factor in the top 5 percent of features, iterative RELIEF does so in ~15 percent of instances.

features. In genome-wide studies, there is a need to develop methods to handle markers in chunks, e.g., grouping SNPs in genetic pathways. Third, insufficient attention has been paid to target selection. The difficulty there will be to identify situations where, e.g., affected targets prove more infor-mative than unaffected ones (or a combination of both). Fourth, if iterative variants of RELIEF show promise, the tuning parameters must be carefully selected, or performance will be seriously degraded. To make these methods applicable in a large scale study (or in a simulation studies), there is a need for more efficient automatic parameter tuning. Finally, having conducted reviews of numerous articles, we have found, with a few notable excep-tions, that obtaining the code for comparison purposes is not always trivial. This suggests the need for a standard and efficient library that will allow such method comparison.

That said, if RELIEF is clearly not a panacea (high power for the example we used would have been suspicious), it does show promise as an additional tool for the study of gene-environment interactions.

References

1. Park, H., & Kwon, H. C. (2007). Extended relief algorithms in instance-based feature filtering. In *Sixth International Conference on Advanced Language Processing and Web Information Technology*, 123–128. IEEE.

2. Amjady, N., Dareepour, A., & Keynia, F. (2010). Day-ahead electricity price forecasting by modified relief algorithm and hybrid neural network. *IET Generation, Transmission & Distribution, 4*(3), 432–444.

3. Wang, Y., Makedon, F., Ford, J., & Pearlman, J. (2005). HykGene: A hybrid approach for selecting marker genes for phenotype classification using microarray gene expression data. *Bioinformatics, 21*(8), 1530–1537.

4. Wang, Y., & Makedon, F. S. Application of Relief-F feature filtering algorithm to selecting informative genes for cancer classification using microarray data. In *Computational Systems Bioinformatics Conference Proceedings* 2004. IEEE.

5. Greene, C. S., Penrod, N. M, Kiralis, J., & Moore, J. H. (2009). Spatially uniform Relieff (SURF) for computationally-efficient filtering of gene-gene interactions. *Bio-Data Mining, 2*(1), 5.

6. Cordell, H. J. (2009). Detecting gene-gene interactions that underlie human diseases. *Nature Reviews. Genetics, 10*(6), 392–404.

7. Kira, K., & Rendell, L. (1992a). A practical approach to feature selection. In *Proceedings of the 1992 Conference on Machine Learning*. Morgan Kaufmann.

8. Kira, K., & Rendell, L. (1992b). The feature selection problem: Traditional methods and a new algorithm. In *Tenth National Conference on Artificial Intelligence*. AAAI Press/MIT Press.

9. Pan, Q., Hu, T., & Moore, J. H. (2013). Epistasis, complexity, and multifactor dimensionality reduction. *Methods in Molecular Biology, 1019*, 465–477.

10. He, H., Oetting, W. S., Brott, M. J., & Basu, S. (2009). Power of multifactor dimensionality reduction and penalized logistic regression for detecting gene-gene interaction in a case-control study. *BMC Medical Genetics, 10*, 127.

11. He, H., Oetting, W. S., Brott, M. J., & Basu, S. (2010). Pair-wise multifactor dimensionality reduction method to detect gene-gene interactions in a case-control study. *Human Heredity, 69*(1), 60–70.

12. Davis, N. A., Lareau, C. A., White, B. C., Pandey, A., Wiley, G., Montgomery, C. G., et al. (2013). Encore: Genetic Association Interaction Network centrality pipeline and application to SLE exome data. *Genetic Epidemiology, 37*(6), 614–621.

13. Moore, J. H., Gilbert, J. C., Tsai, C. T., Chiang, F. T., Holden, T., Barney, N., White, B. C. (2006). A flexible computational framework for detecting, characterizing, and interpreting statistical patterns of epistasis in genetic studies of human disease susceptibility. *Journal of Theoretical Biology, 241*(2), 252–261.

14. Bins, J., & Draper, B. (2002). Evaluating feature relevance: Reducing bias in relief. In *Joint Conference on Information Sciences*, 757–760. Durham, NC.

15. Hastie, T., Tibshirani, R., & Friedman, J. (2010). *The Elements of Statistical Learning: Data Mining, Inference, and Prediction*. Springer.

16. Wu, T. T., Chen, Y. F., Hastie, T., Sobel, E., & Lange, K. (2009). Genome-wide association analysis by lasso penalized logistic regression. *Bioinformatics, 25* (6), 714–721.

17. Yang, J., & Li, Y. (2006). Orthogonal relief algorithm for feature selection. *Lecture Notes in Computer Science, 4113*, 227–234.

18. Greene, C. S., Himmelstein, D. S., Nelson, H. H., Kelsey, K. T., Williams, S. M., Andrew, A. S., et al. (2010). Enabling personal genomics with an explicit test of epistasis. *Pacific Symposium on Biocomputing*, 327–336.

19. Sun, Y., Todorovic, S., & Goodison, S. (2010). Local-learning-based feature selection for high-dimensional data analysis. *IEEE Transactions on Pattern Analysis and Machine Intelligence, 32*(9), 1610–1626.

20. McKinney, B. A., White, B. C., Grill, D. E., Li, P. W., Kennedy, R. B., Poland, G. A., & Oberg, A. L. (2013). ReliefSeq: A gene-wise adaptive-K nearest-neighbor feature selection tool for finding gene-gene interactions and main effects in mRNA-Seq gene expression data. *PLoS One, 8*(12), e81527.

21. Robnik-Šikonja, M., Savicky, P., & Alao, J. A. (2015). *COARElearn: Classification, Regression and Feature Evaluation*. Available from: https://cran.r-project.org/web/packages/COARElearn/index.html.

22. Karney, C. F. F. (2014). *RandomLib: Random number library*. Available from: http://randomlib.sourceforge.net/.

23. King, D. (2009). Dlib-ml: A Machine Learning Toolkit. *Journal of Machine Learning Research, 10*, 1755–1758.

24. Stokes, M. E., & Visweswaran, S. (2012). Application of a spatially-weighted Relief algorithm for ranking genetic predictors of disease. *BioData Mining, 5*, 20–31.

25. Robnik-Šikonja, M., & Kononenko, I. (2003). Theoretical and empirical analysis of ReliefF and RReliefF. *Machine Learning, 53*, 23-69.

26. Bachrach-Gilad, R., Navot, A., & Tishby, N. (2004). Margin based feature selection—Theory and algorithms. In *21st International Conference on Machine Learning*. Banff, Canada.

27. Sun, Y., & Li, J. (2006). Iterative RELIEF for FeatureWeighting. *Proceedings of the 23rd International Conference on Machine Learning*, 913–920.

28. Kononenko, I., Simec, E., & Robnik-Šikonja, M. (1997). Overcoming the myopia of inductive learning algorithms with RELIEFF. *Applied Intelligence*, *7*, 39–55.

29. Yang, P., Ho, J. W. K., Zomaya, A. Y., & Zhou, B. B. (2010). A genetic ensemble approach for gene-gene interaction identification. *BMC Bioinformatics*, *11*, 524.

30. Draper, B., Kaito, C., & Bins, J. (2003). Iterative relief. In *Workshop on Learning in Computer Vision and Pattern Recognition*. Madison, WI.

31. Greene, C. S., et al. (2010). The informative extremes: Using both nearest and farthest individuals can improve relief algorithms in the domain of human genetics. In *Evolutionary Computation, Machine Learning and Data Mining in Bioinformatics, 8th European Conference, EvoBIO 2010*, 182–193. Istanbul.

32. Moore, J.H. & White, B.C. (2007). Tuning ReliefF for genome-wide genetic analysis. *Lecture Notes in Computer Science*, *4447*, 166–175.

33. Zou, H., & Hastie, T. (2005). Regularization and variable selection via the elastic net. *Journal of the Royal Statistical Society. Series B. Methodological*, *67*, 301–320.

34. Byrd, R. H., Lu, P. H., & Nocedal, J. (1995). A limited memory algorithm for bound constrained optimization. *SIAM Journal on Scientific and Statistical Computing*, *16*, 1190–1208.

35. Robnik-Šikonja, M., & Kononenko, I. (1997). An adaptation of Relief for attribute estimation in regression. *Machine Learning: Proceedings of the Fourteenth International Conference (ICML'97)*, 296–304.

36. Beretta, L., Santaniello, A., van Riel, P. L. C. M., Coenen, M. J. H., & Scorza, R. (2010). Survival dimensionality reduction (SDR): Development and clinical application of an innovative approach to detect epistasis in presence of right-censored data. *BMC Bioinformatics*, *11*, 416.

37. McKinney, B. A., Crowe, J. Jr., Guo, J., & Tian, D. (2009). Capturing the spectrum of interaction effects in genetic association studies by simulated evaporative cooling network analysis. *PLOS Genetics*, *5*(3), e1000432.

38. McKinney, B. A., Reif, D. M., White, B. C., Crowe, J. E., Jr., & Moore, J. H. (2007). Evaporative cooling feature selection for genotypic data involving interactions. *Bioinformatics*, *23*(16), 2113–2120.

7 Gene × Environment Interaction in Obesity: The Contribution of Randomized Clinical Trials

Jeanne M. McCaffery and Caroline Y. Doyle

It is widely accepted that gene × environment interaction contributes to growth and development. Yet solid evidence of statistical or biologic gene × environment (G × E) interaction in human studies remains rare. An emerging exception is the field of obesity, in which the contributions of both genetic and environmental factors are well documented, examples of G × E in epidemiologic studies withstand the test of replication, and genetic predictors of success with environmental intervention are increasingly reported.

Understanding the role of genetic and environmental factors and their interaction in obesity is of substantial public health importance. Currently, 64 percent of the US population is estimated to be overweight or obese (body mass index $\geq$ 25 kg/m^2) [1], representing a dramatic increase in prevalence over the past two decades. As a result, millions of Americans now suffer weight-related health complications, including cardiovascular disease, diabetes, and certain cancers [2–4]. Indeed, the progress in reducing cardiovascular morbidity and mortality achieved through reductions in cigarette smoking and control of blood pressure and cholesterol levels is being offset by rises in obesity and the associated dysglycemia [5].

With regard to genetic influence, body weight is heritable with twin studies estimating that genetic factors account for between 40 and 70 percent of the variance in body mass index (BMI) [6, 7]. Common genetic variants associated with obesity have also been identified through genome-wide association studies (GWAS), although the variance attributable to these loci remains small (< 2 percent) [8–10]. Interestingly, many of the genes associated with obesity, including FTO (and *IRX3* [11], *MC4R*, *BDNF*, and *POMC*, are expressed in central nervous system eating pathways (e.g., hypothalamic regions), emphasizing the potential role of energy intake in these associations [10]. The obesity-associated variant in the *FTO* region, in particular, has been associated with a preference for energy-dense food [12], greater consumption of fat and calories [13], and reduced satiety [14]. Many of these loci contributing to common forms of obesity are further located in gene regions for which rare mutations contributing to monogenic forms of obesity have also been described, often contributing to hyperphagia [15,16].

In addition to genetic influence, the dramatic rise in obesity rates over the past thirty years clearly implicates the environment [17]. Easy access to food and fewer opportunities for physical activity stand out as likely causes [18–20] but others have explored the evidence for additional environment influences that correlate with population increases in obesity, such as sleep deprivation, medications, and nonsmoker prevalence [21].

Epidemiologic G × E in Obesity

Given the ample evidence for both genetic and environmental effects, it is not surprising that twin and population-based epidemiologic studies find replicated evidence that genetic risk for obesity can be accentuated or diminished by measured behavior and environment.

Twin Designs

In twin designs, G × E can be defined by differences in the magnitude of genetic variance or heritability associated with a measured environmental attribute, controlling for potential gene-environment correlation [22]. In this regard, it has been demonstrated and replicated that measures of physical activity diminish the heritability of BMI. In male Vietnam-era twins from the United States, heritability of BMI was lower among those reporting high levels of weekly vigorous physical activity compared to those not reporting vigorous activity [23]. Similar results have been documented in male and female young adult twins from Finland [24, 25] and adult twins from Denmark [25].

In a similar twin analyses, dietary protein intake did not appear to alter the heritability of BMI [25]. However, additional twin studies aimed to determine whether the heritability of BMI differs by habitual dietary intake, sleep, various medications, and/or smoking status would be of great interest.

Population-Based Studies of Unrelated Individuals

Conceptually similar to the twin analyses demonstrating attenuation of the heritability of BMI with high levels of physical activity, it has likewise been shown the impact of the obesity-associated *FTO* region on BMI susceptibility varies by level of physical activity [26, 27]. In the first report by Andreasen and colleagues, SNPs in the *FTO* region were associated with obesity only among participants reporting little to no physical activity, with the *FTO* effect being diminished as a function of more vigorous activity [27]. This result has been confirmed for an expanded list of obesity-associated genetic markers by meta-analysis in over 100,000 participants [26].

More recent reports find that genetic risk for obesity is not only diminished by physical activity but exaggerated by TV-watching [28], intake of sugar-sweetened beverages [29], and consumption of fried foods [30]. Importantly, the results were replicated in three large cohorts, the Nurses' Health Study, the Health Professionals Follow-up Study and the Women's Genome Health Study. Such findings validate the notion that environmental exposures, from exercise frequency to dietary food/drink intake, lead to variation in the genetic effects on one's health. Indeed, with the strong pattern of replication, these results lead to the possibility of refining the specific types of environmental exposures that contribute to G × E interactions in obesity, since lack of physical activity, sedentary behavior, and sugar-sweetened beverage and fried food consumption are frequently intercorrelated.

Behavioral Weight Loss

Obesity is also amenable to behavioral change [31, 32]. Randomized, controlled trials of behavioral weight loss programs often produce initial weight losses of 7 percent or more, resulting in clinically important health benefits [33, 34]. For example, in Look AHEAD [35] (figure 7.1), the largest behavioral weight loss trial to date, participants randomized to an intensive lifestyle intervention (ILI), focusing on weight loss through calorie restriction, increased physical activity, and supporting behavioral strategies, lost an average of 8.6 ± 6.9 percent of their initial body weight at year 1 ($n = 2,496$; 97.1 percent follow-up) relative to losses of 0.7 ± 4.8 percent among individuals assigned to the Diabetes Support and Education (DSE) group ($n = 2,463$, 95.7 percent follow-up), who received diabetes support and

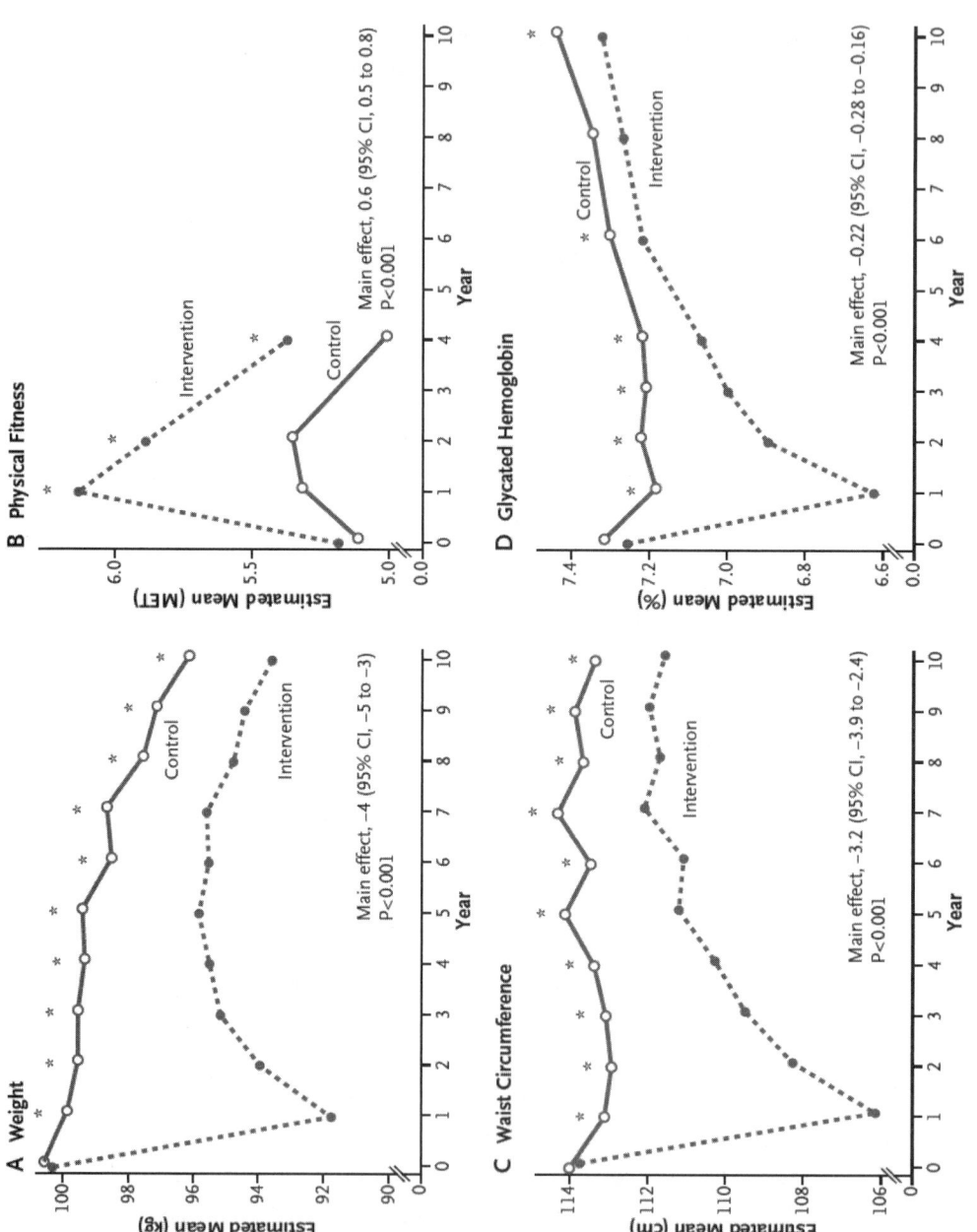

Figure 7.1

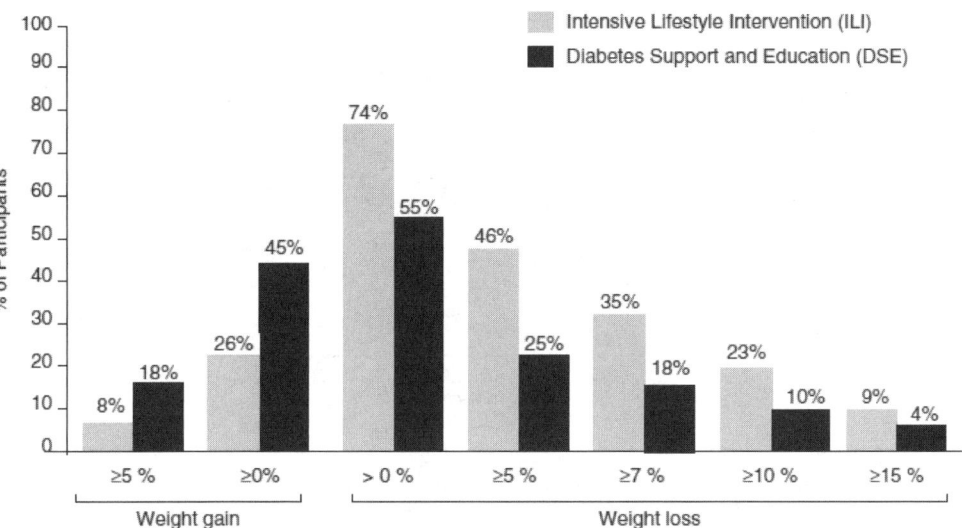

Figure 7.2
Individual differences in weight loss at year 1 in Look AHEAD.

education groups alone [36]. Fully 55.2 percent achieved a 7 percent weight loss in ILI vs. only 7 percent who achieved the goal in DSE.

Partial weight regain is nonetheless common [34, 37]. After the initial weight loss in ILI at one year, participants tended to regain weight although weights, on average, did not return to baseline levels (ILI: 8.6 percent vs. DSE: 0.7 percent at 1 year; ILI: 6.0 percent vs. DSE: 3.5 percent at study end). In Look AHEAD, a gradual weight loss in the DSE group as well as ILI in later years was observed presumably due primarily to joint effects of aging and increasing duration of diabetes which, at least in some observational studies, are associated with weight loss [38].

In addition to the overall trends in ILI and DSE, it is important to note substantial individual differences in response to both ILI and DSE. As can be seen in figure 7.2, although ILI clearly shifted the curve toward greater weight loss and a lower likelihood of weight gain, the distribution of weight change at year 4 in both ILI and DSE were relatively normal with a range of responses from successful to unsuccessful in both conditions. Figure 7.3 further illustrates individual differences in weight regain, here focusing on year 4 weight among individuals who lost 10 percent or more of their initial

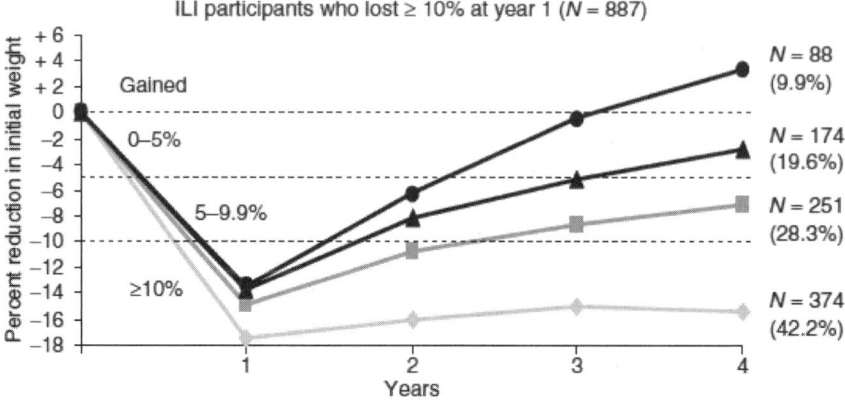

Figure 7.3
Individual differences in weight regain at year 4 among individuals who lost 10 percent or greater of their initial weight in ILI.

weight by year 1 in Look AHEAD. Results range from full maintenance of weight loss to full regain.

Behavioral Predictors of Weight Loss

Clearly, a greater understanding of predictors of individual difference in weight loss and weight maintenance or regain could have important health benefits. Adherence to the intervention is well known to predict better outcomes. In Look AHEAD, for example, physical activity, more frequent attendance to group sessions, and consumption of meal replacements accounted for more than 19 percent of weight loss at year 1 [39]. Readiness to change as assessed by density of entries (greater number of words and numbers) entered into self-monitoring diaries further predicted weight loss at year 1 in Look AHEAD [40].

Genetic Predictors of Weight Loss

Less is known about genetic predictors of ability to lose weight and maintain weight losses despite the public health importance and relevance to G × E interaction of these questions. A single twin study of seven identical twin pairs has tested for heritability of weight loss. Bouchard and colleagues [41] induced weight loss among these identical twins through supervised daily exercise under a regimen of constant daily energy intake. In this highly controlled laboratory setting, participants lost on average roughly 5 kg. Yet, individuals differed substantially in the magnitude of weight loss. The correlation between the identical twins was $r = 0.74$, suggesting that the magnitude of genetic influence on weight loss in this controlled laboratory setting was quite high, perhaps higher than prior reports for obesity per se.

Research has also begun to identify specific genetic markers that predict success with weight loss with behavioral intervention, e.g., [42–44]. Many of these studies present the opportunity to study G × E in the unique context of the RCT design. In the next sections, we will review several methodological advantages of examining G × E in RCTs, followed by a discussion of research on genetic predictors of weight loss in the Look AHEAD trial as an applied example.

The Randomized, Controlled Trial Design in G × E Research

The effective treatments in obesity research afford a strong opportunity to test gene × environment interaction where a behavior change intervention can be randomly assigned and compared to a control condition over time in an RCT design. The most recent AHA/ACC/TOS guidelines recommend—with the highest level of evidence—that overweight and obese individuals participate for six months or more in a comprehensive lifestyle program that assists participants in adhering to a lower-calorie diet and in increasing physical activity through the use of behavioral strategies [45, 46]. For calorie restriction, the report assigns the strong level of evidence for the use of: (1) moderately reduced caloric intake (e.g., 1,200–1,500 kcal/day for women and 1,500–1,800 kcal per day for men; (2) a 500 or 750 kcal per day energy deficit; or (3) evidence-based diets that restrict certain food types producing an energy deficit. For physical activity, increased aerobic physical activity (such as brisk walking) for more than thirty minutes per day most days of the week is recommended. Moreover, behavioral strategies to support success with behavior change are also recommended, including self-monitoring of food intake, physical activity, and weight.

The control arms in weight loss RCTs are often "usual care" or involve a minimal contact condition, such as a newsletter or educational meetings, to minimize attrition. Increasingly, however, control arms match on factors such as participant contact to minimize distinctions between the intervention and control arms.

With the RCT design, it can be argued that participants are randomly assigned to one "environment" (i.e., treatment arm) or another. This randomization of "environment" affords several methodological advantages when addressing G × E hypotheses.

Measurement of Environment

A key challenge to G × E research is defining the environment. Environmental exposures are often difficult to measure objectively, with measurement commonly relying on self-report that can be subject to measurement error, recall bias, and/or lack of precision. Certainly more detailed characterization of the putative exposures in prospective studies using reliable and valid measurement tools and, if possible, objective measurement, can improve the accuracy of measurement of environmental exposure. Technology, such as smart phones, for example, can provide greater precision in measurement of exposures throughout the day instead of relying on self-report assessments often covering weeks, months, or years. Blood levels or environmental sensors can also increase accuracy. But state-of-the-art measurement is frequently time-consuming and expensive. Furthermore, such detailed measures are unlikely to be available in larger cohorts or replications cohorts, with statistical power and replication being key metrics for the evaluation of the contribution of a G × E result.

In RCTs, randomized "environment" is defined through research design with little to no measurement error in treatment allocation. Despite the clarity of research design, it is important to recognize that a treatment arm is quite distinct from what one would anticipate as an "exposure" in G × E research. Treatment arms in behavioral weight loss studies are typically multifaceted, and *any* way in which one treatment arm differs from another contributes to differential effectiveness. In addition to prescriptions for calories, physical activity, and self-monitoring, the interventions are often conducted in group settings with additional therapist contact, creating gradients of social support and expert guidance across interventions and control arms. Treatment fidelity and blinding of assessment staff must be maintained to minimize unintended differences between treatment arms. With these caveats, however, lack of measurement error in environmental exposure is an important advantage of RCTs in G × E research.

Longitudinal Design

RCTs most commonly follow participants over a period to determine effectiveness. Thus, the outcome of interest is measured prospectively, avoiding recall bias. Change in weight is measured objectively by blinded researchers, further enhancing the accuracy of measurement. Moreover, as RCTs typically include repeated measures, longitudinal models can be constructed that minimize the influence of measurement error at any given time point. Last, to the extent that significant interaction is observed, longitudinal mediation models can be constructed to isolate specific processes of change for future intervention development.

Experimental Control of Covariates

Another clear challenge of G × E research in epidemiologic studies is the difficulty in isolating specific aspects of environment's contributing to a gene × environment interaction finding. Environmental risk factors often cluster together creating complexity in delineating whether a given environmental characteristic plays a causal role. Correlated measured environmental characteristics can be statistically controlled in regression models in an attempt to identify one of several correlated variables but the success of this method will depend on the accuracy with which the potentially confounding variables are measured and the magnitude of intercorrelation, which can create unstable statistical models. Unmeasured variables may also account for the association of an environmental characteristic and outcome in G × E research.

The RCT is designed to minimize or eliminate differences in baseline characteristics and exposures across treatment arms. As treatment condition is assigned randomly, typically using a random number generator, it can be assumed that the treatment and control groups do not differ meaningfully on demographic characteristics, and measured or unmeasured environmental characteristics at baseline, an assumption that, for measured variables, can be verified through statistical confirmation.

For this strength to hold, the statistical analysis of outcome data in G × E studies within RCTs must be examined as G × T (gene × treatment arm interaction). Individual behavior, such as whether or not individuals actually changed their diet or physical activity, can subsequently be tested as process variables using mediation analyses, but randomized treatment arm must remain the primary instrumental variable.

In addition, $G \times T$ must be based on *all* participants randomized in the experiment. Differential attrition may occur by treatment arm undermining comparability of participants across treatment arms and biasing results. In weight loss interventions, for example, participants who are less successful in weight loss are more likely to drop out. Drop out may further be biased by demographic, environmental, or behavioral characteristics. In "Completer" analyses, only individuals who completed the intervention are analyzed. To the extent that differential drop out has occurred, any result can no longer be attributable to solely to the randomized intervention. In "Intent to treat" analyses, in contrast, all randomized participants are included in analysis even if lost to follow-up. In weight loss treatment, it is typically assumed that participants who dropped out either have returned to baseline weight or have a weight higher than the last available measurement. As these are assumptions, every effort should be made to minimize attrition.

Lack of G-E Correlation

One potential confounder of $G \times E$ analyses that deserves particular note is gene-environment correlation. In a classic review, Kendler and Baker [47] demonstrated that many of the exposures commonly assumed to be environmental contributors to growth and development reflect, at least in part, genetic influence. Exposures ranging from stressful life events, to parent behavior, to family environment, to social support show at least low to moderate heritability (genetic variance within a population), with a mean heritability of 0.27 across these exposures. It is plausible that gene-environment correlation may be *active* if environmental experiences were selected based upon heritable predispositions. Alternatively, they could be *reactive*, such as when the environment is changed in response to heritable characteristics. Finally, they could also be passive due to the correlation of genetic risk with environments provided by parents who share similar genetic risks.

A recent illustration of the potentially broad-reaching impact of gene-environment correlation comes from the work of Christakis and Fowler [48] who demonstrated that friends share on average the same genetic similarity as fourth cousins despite a lack of recent common ancestors, indicating that people initiate and maintain friendships based on covert genetic similarity or that people who are similar genetically will select similar environments in which they are more likely to meet and become friends.

The epidemic of obesity corresponds to secular changes in food availability, occupational physical activity, and technology, macroenvironmental trends that have occurred across a relatively short time frame (1–2 generations) and are unlikely to reflect solely genetic effects (although they do appear to contribute to G × E).

Nonetheless, many of the environmental exposure thought to contribute to obesity, such as caloric intake and physical activity, have been shown to be heritable [49, 50]. Moreover, several of the loci associated with BMI in GWAS, including *FTO*, *BDNF*, *MC4R*, *SH2B1*, and *POMC* are expressed in the central nervous system—in the feeding centers of the hypothalamus in particular—highlighting a neuronal component in obesity predisposition and the potential for direct genetic effects on eating behavior [51]. Indeed, early studies found links between such loci and distinct dietary patterns such as preference for energy dense food [12], greater consumption of fat and calories [13], and diminished satiety [14], albeit predominantly in children. More recently, the obesity-associated locus within *FTO* was associated with more eating episodes per day, greater total caloric intake, and eating of a greater number of meals and snacks per day, and risk variants within *BDNF* predicted greater total caloric intake and more servings from the dairy and meat, nuts, and beans food groups [51]. Thus, there is strong potential for gene-exposure correlation confounding the gene-environment interaction in obesity.

Designed to eliminate measured and unmeasured biases across treatment arms, RCTs should minimize any gene-environment correlation when analyses are conducted as G × T including all participants randomized. As random variation can occur in participant allocation to treatment arms, it is the nonetheless important to conduct randomization checks for measured genetic and environmental variables.

Successful Intervention

Although G × E interaction can occur in the absence of a direct effect of environment on the outcome, exposures demonstrating consistent and replicated association with the outcome of interest serve as excellent candidate exposures for G × E analyses. Above and beyond association, successful RCTs augment the certainty with which causal attributions can be made, substantially increasing the likelihood that the exposure contributes to the etiology of the outcomes of interest.

Within select RCTs, an exposure may further be shown to have a causal effect on the biology of an outcome, providing a context for understanding the biological mechanisms underpinning a statistical G × E. Behavioral weight loss intervention changes adipose tissue composition, inflammation, and insulin resistance, to name a few. Measuring these biologic pathways provides a context to delineate how behavioral weight loss changes physiology and the pathways through which SNPs and other genetic variants may alter these effects. Thus, there are opportunities for biologic validation of statistical G × E interactions in many RCTs. Incorporating neuroimaging will be of particular interest, since many of the genetic and environmental pathways of obesity are thought to occur in the brain.

Unique Phenotypes of Public Health Importance

Many people are attempting to lose weight, either on their own or using various commercial programs or apps, often based on physician recommendation to maintain or improve their health. With the diversity of methods, differences in efficacy, and little objective assessment, intentional weight loss is a very difficult phenotype to characterize on a population level. One could measure change in weight in an epidemiological study but it is difficult to determine whether weight loss reflects intention or potentially medical illness or frailty. A snapshot of weight at any given time further reflects genetic and environmental influence on weight gain, as well as weight loss and regain, each possibly with distinct genetic and environmental contributors. Behavioral weight loss RCTs provide a unique opportunity to delineate the multiple phenotypes and identify genetic and environmental influence.

Clinical Application

A frequent justification of G × E research is that interventions may be developed to reduce genetic risk. In G × E studies within RCTs, a well-characterized intervention has already been designed based on a hypothesized impact on the outcome of interest, greatly facilitating translation to clinical application. G × E studies can build upon this platform to inform the outcomes of clinical trials and potentially serve as a basis to optimize results.

Many individuals are not successful even with initial weight loss, yet recommendations for caloric restriction and increased physical activity have traditionally been "one size fits all." Identifying specific genes that predict individual differences in weight loss in response to behavioral intervention may help identify individuals with differing patterns of weight loss and guide a tailoring of behavioral strategies to optimize weight loss in more individuals.

Models of Gene × Treatment Interaction

Two primary types of G × T interaction effects can occur with obesity RCTs.

Qualitative Interaction 1: Gene × Treatment Arm Interaction for Loci Predictive of Weight Gain

One interaction potentially detected in weight loss RCTs relates to loci associated with weight gain, but not weight loss per se. Here, we conceptualize high caloric intake and low levels of physical activity as permissive factors necessary for the expression of genetic vulnerability to obesity. In the absence of the permissive behaviors, genetic vulnerability to obesity would not be expressed. This scenario is similar to the epidemiologic observation that rs9939609 in *FTO* is associated with BMI among those reporting low levels of physical activity, but that the association is not seen among those reporting high levels of physical activity [27]. Within a weight loss trial, this type of interaction could occur if the intervention were sufficient to overshadow the effects of genetic vulnerability, resulting in a significant effect of genotype on weight change in a no-treatment control arm but little to no significant effect of genotype in the behavioral weight loss arm (figure 7.4).

Qualitative Interaction 2: Gene × Behavior Interaction for Loci Predictive of Weight Loss

Available twin studies demonstrate that genetic factors influence ability to lose weight even given the same levels of caloric restriction or physical activity. Model 7.2 (figure 7.5) depicts effects of genetic markers that influence weight loss only (effects seen in the treatment arm but not the control arm). In the prior literature, for example, *PPARG* Pro12Ala homozygotes lost more weight than Pro carriers, a genetic difference that was limited to the lifestyle intervention arm in the Finnish Diabetes Prevention Program [52] and the US Diabetes Prevention Program [53].

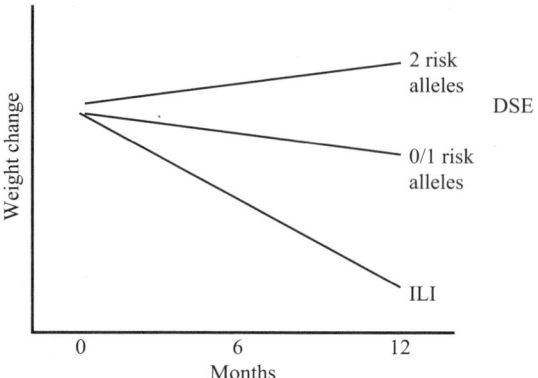

Recessive gene x treatment interaction in Model 1

Figure 7.4
Model of qualitative interaction 7.1.

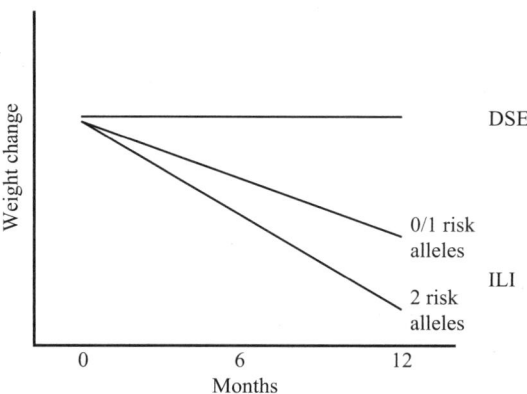

Recessive gene x treatment interaction in Model 2

Figure 7.5
Model of qualitative interaction 7.2.

Quantitative (Crossover) G × T Interaction

It is plausible that one or more loci may impact weight loss in both the intervention and control arms resulting in combinations of models 7.1 and 7.2. For example, a SNP associated with weight gain in a control arm may be associated with more rapid weight loss in the intervention arm. This scenario could result in genetic associations within both arms but in opposite directions and potentially nonsignificant within each arm.

No G × T Interaction

It is also possible that genetic regions that promote weight gain also predict resistance to weight loss. This combination of two genetic associations is unlikely to result in a gene × treatment arm interaction in the context of the clinical trial, as we would predict that the loci would have similar effects in the intervention and control arms. It would nonetheless identify genetic regions associated with both weight gain and resistance to weight loss, effects of key public health importance.

It is also worth noting that no genetic loci may predict change in weight in either the treatment or control condition, or interact with treatment arm. This may reflect a "true negative" result, in which the results of the G × T experiment reflect the underlying truth that no genetic loci are associated with treatment response. Alternatively, a lack of association in this experiment may reflect a "false negative" due to lack of statistical power and small effect size as the control of most potential confounders is implicit in the RCT design.

Application of G × T in the Look AHEAD Study

To illustrate the implementation of G × T models in the context of behavioral weight loss RCTs, we review our prior research from the Look AHEAD Genetic Ancillary Studies.

The Look AHEAD Study

The primary goal of the Look AHEAD Study was to determine whether weight loss achieved through an intensive lifestyle intervention (ILI), including diet and physical activity, can reduce cardiovascular morbidity and mortality among persons with type 2 diabetes, compared to diabetes support and education (DSE) alone [34, 35, 37]. The cohort included 5,145 participants who were overweight (> 25 kg/m^2 or > 27 kg/m^2 if on insulin), had type 2 diabetes and were between the ages of 45–74 at baseline; 30 percent reported racial or ethnic minority ancestry.

Participants were randomly assigned to ILI or DSE. Both the ILI and DSE groups were provided education on diabetes and cardiovascular risk factors. In addition, ILI was designed to produce an average of 7 percent weight loss and maintain the weight losses through lifestyle intervention. Participants in this arm received a goal of >10 percent weight loss in order to increase the likelihood of achieving an average of 7 percent weight loss across participants. The ILI included one individual and three group meetings per month for six months followed by a minimum of one face-to-face contact per month through four years and a minimum of biannual contact through study end. These sessions focused on behavioral weight loss strategies, such as self-monitoring, goal setting, and stimulus control, to achieve and maintain weight loss. ILI also included a "toolbox" of advanced behavioral techniques and use of a weight loss medication (orlistat) for participants having difficulty meeting weight or physical activity goals. The DSE group received the option of attending three sessions per year on nutrition, physical activity, and social support with no explicit weight loss goals.

The intervention component of the trial was stopped early on the basis of a futility analysis for the primary outcome, cardiovascular disease morbidity and mortality, when the median follow-up was 9.6 years. Weight loss was greater in the intervention group than in the control group throughout the study (8.6 percent vs. 0.7 percent at 1 year; 6.0 percent vs. 3.5 percent at study end). ILI also produced greater reductions in glycated hemoglobin and greater initial improvements in fitness and all cardiovascular risk factors, except for low-density-lipoprotein cholesterol levels. The primary outcome occurred in 403 patients in the intervention group and in 418 in the control group (1.83 and 1.92 events per 100 person-years, respectively; hazard ratio in the intervention group, 0.95; 95 percent confidence interval, 0.83 to 1.09; P = 0.51) [37].

Genotyping Platforms

A primary goal of the Look AHEAD Genetic Ancillary Studies was to determine whether single nucleotide polymorphisms (SNPs) associated with obesity in genome-wide association studies (GWAS) predicted the magnitude of change in weight in response to weight loss intervention. Genotyping was conducted with two chips, the *Illumina CARe iSelect* (IBC) chip, a gene-centric 50,000 SNP array designed to assess relevant loci across a range of cardiovascular, metabolic and inflammatory syndromes [54]; and the Metabochip, a custom Illumina iSELECT genotyping array designed to test ~200,000 SNPs identified through GWAS meta-analyses for metabolic and CVD traits, focusing primarily on MI, CAD, LDL- and HDL-cholesterol, triglycerides, BMI, and systolic and diastolic BP [55]. Effective sample sizes taking into account genetic consent and quality control of genotyping data were 3,899 and 4,045 for the IBC chip and Metabochip respectively.

Statistical Methods

The goal of the first study was to determine the association between genetic loci previously associated with obesity in GWAS and available on the *Illumina CARe iSelect* IBC chip (13 SNPs within or in the region of *FTO*, *SH2B1*, *MC4R*, *BDNF*, *TNNI3K*, *MTIF3*, *MAP2K5*, *QPCTL/GIPR*, *and TFAP2B*) with weight loss at 1 year in ILI and DSE, and weight regain from year 1 to 4 among those who lost 3 percent or more of their initial weight at year 1. To control for admixed study population, all IBC SNPs were examined by principal component analysis (PCA), indicating that the majority of the variance among the Look AHEAD cohort was accounted for by the first two principal components, which agreed with self-reported ethnicity.

Longitudinal linear mixed models were used to model the effect of SNP on weight change by treatment arm over time. Because baseline weight as well as treatment response can be associated with the SNPs, baseline was modeled as the first time point in longitudinal analyses as recommended by McArdle and Whitcomb [56]. Within this model, differential SNP effects on year 1 weight change or by treatment arm are detected through SNP (0, 1, or 2 copies of the minor allele) × time (baseline, year 1) × treatment arm (ILI, DSE) interaction. An additive genetic model was used for all markers, with genotype coded by the number of minor alleles. Longitudinal outcomes were additionally adjusted for age, gender, study site, and the first two ancestry informative marker principal components.

We further examined the extent to which the genetic markers or G × T interaction predicted weight regain at year 4 among those who lost 3 percent or more of their initial weight at year 1. The same covariates were employed as above, with the addition of year 1 weight. Family-wise error rate was maintained at 0.05 via Sidak's adjustment for multiplicity by declaring as statistically significant only those markers with a nominal significance level of 0.05/10 = 0.005, with 10 reflecting the number of principal components derived from the genotypic correlation matrix [57].

GxT with Loci Previously Associated with Obesity

Obesity risk alleles in *FTO*, *SH2B1*, and *QPCTL/GIPR* regions predicted baseline weight in directions consistent with prior research. Risk alleles for the markers in these three genes were associated with elevated baseline weight of 1.01–1.29 kg per copy. To our surprise, no SNPs were significantly associated with the magnitude of weight change in either ILI or DSE or interacted with treatment arm in predicting the degree of weight change at year 1, indicating that the obesity-associated alleles available in this data set did not deleteriously impact ability to lose weight with behavioral efforts in this largest study of behavioral weight loss. Nor did they influence degree of naturalistic change in weight over the course one year, at least in this study.

However, the obesity risk (A) allele at *FTO* rs3751812 was significantly associated with weight regain in DSE (1.559 kg per risk allele, $p = 0.005$), but not ILI (-0.092 kg per risk allele, $p = 0.761$), resulting in a nominally significant SNP × treatment arm interaction ($p = 0.009$). This *FTO* association was predominantly attributable to weight regain in the control arm and not the intervention arm, suggesting that obesity associated alleles in the *FTO* region may promote weight regain in the permissive environment in DSE, an effect that does not occur with continued lifestyle treatment in ILI.

Discovery G × T

We next leveraged all of the SNPs available on the IBC chip to determine if other loci might predict weight loss or weight regain. The same sample, outcomes and statistical methods were used as documented above but we examined all autosomal markers with minor allele frequency (MAF) < .05, including 31,692 SNPs. Using the effective number of uncorrelated markers of 17,254 after lindage disequilibrium correction [57], we calculated the chipwide significance threshold at $P = 2.97E-06$. We further used a false discovery rate (FDR) approach [58] to guide our reporting of *suggestive* (FDR < 10 percent) associations for further replication calculated using the *Q-value* package of Dabney and Storey [59].

Weight Loss A Manhattan plot depicting association of the full set of SNP markers with year 1 weight change in ILI is presented in figure 7.6.

The association of two loci with year 1 weight change in the ILI group exceeded chipwide significance after correcting for chipwide multiple comparisons ($p < 2.96E-06$). One intronic locus represented by two SNPs on chromosome 2, rs484066 and rs569805, within *ABCB11*, showed the strongest association with year 1 weight loss. These SNPs were associated with a 1.16 and 1.24 kg higher weight per minor allele at year 1, respectively, suggesting that the minor allele was associated with resistance to weight loss. A third chromosome 18 SNP, rs17069904, within *TNFRSF11A*, or *RANK*,

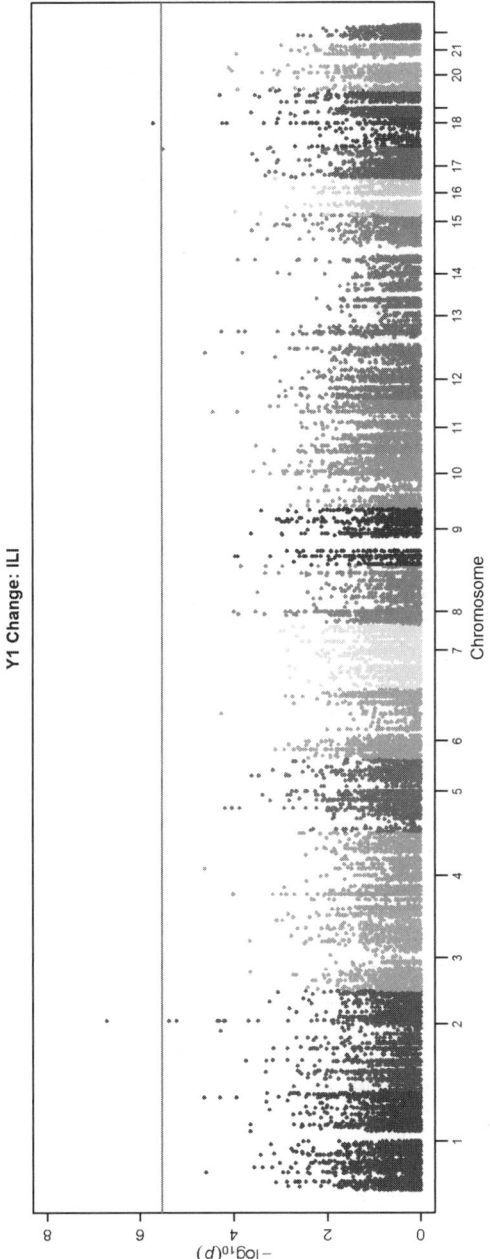

Figure 7.6

Manhattan plot of IBC chip association with year 1 change in weight in ILI.

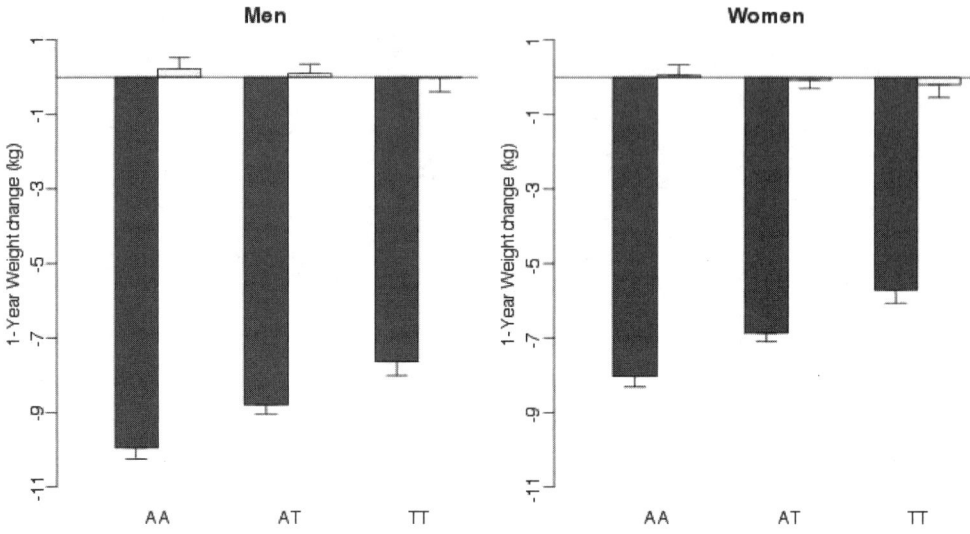

Figure 7.7
Association of *ABCB11* rs484066 with year 1 weight loss in ILI.

also achieved chipwide significance. This SNP was associated with a 1.70 kg lower weight per allele at year 1, suggesting that the minor allele was associated with greater weight loss. These SNPs had no significant effects in DSE (p > 0.51). The resulting SNP × treatment arm interactions for the lead SNP at each locus were *ABCB11* rs484066 interaction $p = 3.98E\text{-}05$; and *TNFRSF11A* rs17069904 interaction: $p = 1.57E\text{-}04$. The associations of the *ABCB11* rs484066 and *TNFRSF11A* rs17069904 with weight loss are depicted in figures 7.7 and 7.8, respectively.

An additional 38 SNPs showed suggestive association with year 1 weight change (FDR $q < 0.10$) but no SNP showed suggestive evidence for association with year 1 weight change in DSE (FDR $q > 0.99$), year 1 weight loss as averaged across treatment arms (FDR $q > 0.99$) or SNP × treatment arm interaction (FDR q > 0.20).

ABCB11, or ATP-binding cassette, subfamily B, member 11, also called bile salt export pump (*BSEP*), is the primary mediator of bile salt secretion across the canalicular membrane and plays a critical role in absorption of dietary fat from the gut and counter transport of hepatic cholesterol from the liver to the intestine for elimination [60]. *ABCB11* knockout mice have smaller body size than wild-type litter mates [61], while overexpression of

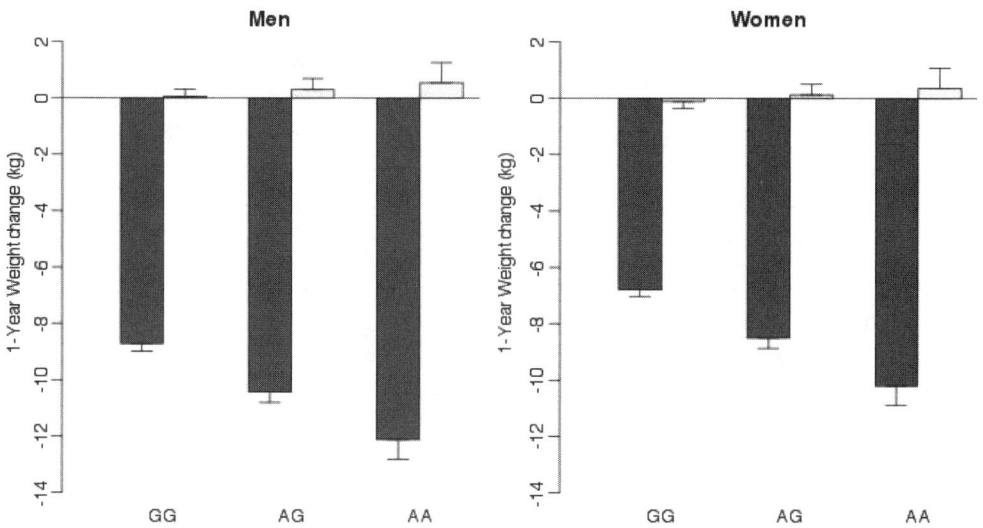

Figure 7.8
Association of *TNFRSF11A* rs17069904 with year 1 weight loss in ILI.

ABCB11 in mice leads to greater fat absorption from the intestine, more rapid weight gain, and a reduction in energy expenditure in response to a high-fat diet but not on a control diet [62]. *ABCB11* variation has also been implicated in obesity in candidate gene studies [63], and fasting HDL and glucose in GWAS [64]. It is of note that this locus is also proximal to glucose-6-phosphatase, catalytic, 2 (*G6PC2*), associated with fasting glucose in genome-wide association studies [65].

The *TNFRSF11A*, or as more commonly known *RANK*, is a member of the TNF family of genes that with osteoprotegerin (OPG) form a signaling network that regulates bone mineral density [66]. Links between change in weight and bone mass density are well described and highlight reciprocal pathways of influence in adipose and bone [66]. *RANK* has been associated with percentage fat mass in animal linkage studies [67] and BMI in humans [68].

Weight Regain No suggestive associations were observed within ILI (FDR $q > 0.99$) or DSE (FDR $q > 0.10$) for change in weight from year 1 to year 4 among individuals who lost ≥ 3 percent of their initial weight. Five SNPs in three loci also showed suggestive evidence of a main effect on weight change across treatment arms (FDR $q < 0.10$). The closest genes for these SNPs were *FOXP1*, *MMP13*, and *TGFBR3*. Finally, 4 SNPs showed suggestive G × T interaction (FDR $q < 0.10$), including two within *GRB2* and one within *TIMP3*. It is notable that several of the SNP associations with weight regain phenotype occur in biologic pathways integrally involved in structural remodeling of adipose tissue and fibrosis accumulation with increasing fat mass [69].

Bird's-Eye View

Stepping back and reviewing our G × T results to date, several patterns emerge. First, SNPs associated with obesity in GWAS studies do not appear to have a strong effect on ability to lose weight, indicating that individuals can successfully lose weight regardless of their genotypes at the GWAS obesity SNPs that have been analyzed to date. This is an important public health message. On the contrary, we do see some association of the *FTO* region, and perhaps *BDNF* rs6265, with weight regain. The *FTO* effect occurred particularly in the control arm, suggesting ongoing intervention may diminish this effect in a manner similar to prior reports of G × E in epidemiologic studies. Associations between *FTO* and *BDNF* are also consistent with other work in Look AHEAD showing that obesity associated alleles at these loci are also associated with greater daily caloric intake at baseline [51]. Thus, *FTO* and *BDNF* may bias eating habits contributing to weight gain, or regain, but, for *FTO*, the regain effect appears to be diminished by continued weight loss intervention resulting in at least a nominal G × T interaction.

Chipwide analyses predominantly revealed potentially new information about weight loss and regain. For weight loss, for example, a substantial number of suggestive effects were observed (41 SNPs, FDR $q < 0.10$), including three SNPs in two loci with statistical associations exceeding the experiment-wise level for significance but with no known effect on weight in either cross-sectional studies or prospective studies of weight change.

Taken together, our results support the idea that genetic evaluation of phenotypes under distinct environmental conditions may yield novel insights for the outcomes of interest. It is often argued that the sample sizes required to detect G × E will be very large, and with good reason. However, it is possible that as yet undetected large effect sizes may occur for novel phenotypes. For instance, a SNP in the region of the glutamate synthase gene (*GLUL*) was identified as associated with cardiovascular disease only among individuals with diabetes [70]. The effect size for this single SNP was larger than all prior SNPs in GWAS combined, accounting for roughly 10 percent of the variance in CVD.

Our research also suggests that perturbing the environment with an intervention may lead to novel genetic insights. As obesity typically develops over a lifetime with cumulative environmental and behavioral risk, cross-sectional G × E interaction may be difficult to capture. Moreover, prevailing environmental conditions promote similar behavior throughout the population, potentially resulting in little variability in environmental or behavioral exposure to detect G × E despite joint genetic and environmental factors contributing to an outcomes such as obesity. Perturbing the environment through intervention or identifying naturalistic periods of dynamic environmental or behavioral change, such as weight regain and pregnancy, may create opportunities to detect novel G × E.

Another implication of our work is a distinction between G × T and clinical significance. Despite strong genetic effects of *ABCB11* rs484066 reaching experiment-wide statistical significance in ILI and little to no genetic effect in DSE, the SNP × treatment arm interaction did not approach statistical significance. This example highlights the extent to which main genetic effects even within half the sample size may have greater statistical power than evaluating a G × T interaction in twice the sample size.

Whether or not a genetic marker is clinically relevant depends upon the magnitude of effect in the treatment arm, often with a lack of effect hypothesized in a control arm. This type of qualitative interaction lacks power relative to a quantitative interaction in which the direction of genetic effect is opposite across treatment arms but often with a smaller magnitude of effect in the treatment arm. Thus, if "clinical significance" of genetic effects were predicated on significance of the G × T interaction alone, genetic markers with the strongest effects in the treatment arm might be overlooked. Moreover, clinically significant effects may occur in the absence of G × T such as if genetic loci that predicted resistance to weight loss also predicted weight gain with similar effects in a treatment and control arm.

Last, it must be recognized that the gold standard for evaluating findings from discovery analyses is replication, which can be particularly challenging in G × E and G × T. In G × T research, identifying trials with a similar intervention, target population, and outcome measures is challenging as each trial is designed to address a unique question. Nonetheless, a future direction of the present research is to partner with the Diabetes Prevention Program (DPP) [31, 32], a randomized, controlled trial designed to determine whether lifestyle intervention or the diabetes medication metformin reduced incidence of type 2 diabetes among overweight or obese participants with impaired glucose tolerance. Hence, although challenging, there may be innovative solutions to systematically pursue the study of G × E relationships within G × T designs that will advance science related to obesity.

References

1. Ogden, C. L., Carroll, M. D., Curtin, L. R., McDowell, M. A., Tabak, C. J., & Flegal, K. M. (2006). Prevalence of overweight and obesity in the United States, 1999–2004. *Journal of the American Medical Association, 295*(13), 1549–1555.

2. Colditz, G. A. (1992). Economic costs of obesity. *American Journal of Clinical Nutrition, 55*, 503–507.

3. Pi Sunyer, F. X. (1993). Medical hazards of obesity. *Annals of Internal Medicine, 119*, 655–660.

4. VanItallie, T. B., & Abraham, S. (1985). Some hazards of obesity and its treatment. In J. Hirsch & T. B. VanItallie (Eds.), *Recent Advances in Obesity Research* (pp. 1–19). Libbey.

5. Stewart, S. T., Cutler, D. M., & Rosen, A. B. (2009). Forecasting the effects of obesity and smoking on U.S. life expectancy. *New England Journal of Medicine, 361*(23), 2252–2260.

6. Maes, H. H., Neale, M. C., & Eaves, L. J. (1997). Genetic and environmental factors in relative body weight and human adiposity. *Behavior Genetics, 27*(4), 325–351.

7. Stunkard, A. J., Foch, T. T., & Hrubec, Z. (1986). A twin study of human obesity. *Journal of the American Medical Association, 256*(1), 51–54.

8. Speliotes, E. K., Willer, C. J., Berndt, S. I., Monda, K. L., Thorleifsson, G., Jackson, A. U., et al. (2010). Association analyses of 249,796 individuals reveal 18 new loci associated with body mass index. *Nature Genetics, 42*(11), 937–948.

9. Thorleifsson, G., Walters, G. B., Gudbjartsson, D. F., Steinthorsdottir, V., Sulem, P., Helgadottir, A., et al. (2009). Genome-wide association yields new sequence variants at seven loci that associate with measures of obesity. *Nature Genetics, 41*(1), 18–24.

10. Willer, C. J., Speliotes, E. K., Loos, R. J., Li, S., Lindgren, C. M., Heid, I. M., et al. (2009). Six new loci associated with body mass index highlight a neuronal influence on body weight regulation. *Nature Genetics*, *41*(1), 25–34.

11. Smemo, S., Tena, J. J., Kim, K. H., Gamazon, E. R., Sakabe, N. J., Gomez-Marin, C., et al. (2014). Obesity-associated variants within FTO form long-range functional connections with IRX3. *Nature*, *507*(7492), 371–375.

12. Cecil, J. E., Tavendale, R., Watt, P., Hetherington, M. M., & Palmer, C. N. (2008). An obesity-associated FTO gene variant and increased energy intake in children. *New England Journal of Medicine*, *359*(24), 2558–2566.

13. Timpson, N. J., Emmett, P. M., Frayling, T. M., Rogers, I., Hattersley, A. T., McCarthy, M. I., et al. (2008). The fat mass- and obesity-associated locus and dietary intake in children. *American Journal of Clinical Nutrition*, *88*(4), 971–978.

14. Wardle, J., Carnell, S., Haworth, C. M., Farooqi, I. S., O'Rahilly, S., & Plomin, R. (2008). Obesity associated genetic variation in FTO is associated with diminished satiety. *Journal of Clinical Endocrinology and Metabolism*, *93*(9), 3640–3643.

15. Farooqi, I. S., Keogh, J. M., Yeo, G. S., Lank, E. J., Cheetham, T., & O'Rahilly, S. (2003). Clinical spectrum of obesity and mutations in the melanocortin 4 receptor gene. *New England Journal of Medicine*, *348*(12), 1085–1095.

16. Gray, J., Yeo, G. S., Cox, J. J., Morton, J., Adlam, A. L., Keogh, J. M., et al. (2006). Hyperphagia, severe obesity, impaired cognitive function, and hyperactivity associated with functional loss of one copy of the brain-derived neurotrophic factor (*BDNF*) gene. *Diabetes*, *55*(12), 3366–3371.

17. McAllister, E. J., Dhurandhar, N. V., Keith, S. W., Aronne, L. J., Barger, J., Baskin, M., et al. (2009). Ten putative contributors to the obesity epidemic. *Critical Reviews in Food Science and Nutrition*, *49*(10), 868–913.

18. Sothern, M. S. (2004). Obesity prevention in children: Physical activity and nutrition. *Nutrition*, *20*(7–8), 704–708.

19. Waxman, A. (2004). WHO global strategy on diet, physical activity and health. *Food and Nutrition Bulletin*, *25*(3), 292–302.

20. Weinsier, R. L., Hunter, G. R., Heini, A. F., Goran, M. I., & Sell, S. M. (1998). The etiology of obesity: Relative contribution of metabolic factors, diet, and physical activity. *American Journal of Medicine*, *105*(2), 145–150.

21. Keith, S. W., Redden, D. T., Katzmarzyk, P. T., Boggiano, M. M., Hanlon, E. C., Benca, R. M., et al. (2006). Putative contributors to the secular increase in obesity: Exploring the roads less traveled. *International Journal of Obesity*, *30*(11), 1585–1594.

22. Purcell, S. (2002). Variance components models for gene-environment interaction in twin analysis. *Twin Research*, *5*(6), 554–571.

23. McCaffery, J. M., Papandonatos, G. D., Bond, D. S., Lyons, M. J., & Wing, R. R. (2009). Gene × environment interaction of vigorous exercise and body mass index among male Vietnam-era twins. *American Journal of Clinical Nutrition*, *89*(4), 1011–1018.

24. Mustelin, L., Silventoinen, K., Pietilainen, K., Rissanen, A., & Kaprio, J. (2009). Physical activity reduces the influence of genetic effects on BMI and waist circumference: A study in young adult twins. *International Journal of Obesity*, *33*(1), 29–36.

25. Silventoinen, K., Hasselbalch, A. L., Lallukka, T., Bogl, L., Pietilainen, K. H., Heitmann, B. L., et al. (2009). Modification effects of physical activity and protein intake on heritability of body size and composition. *American Journal of Clinical Nutrition*, *90*(4), 1096–1103.

26. Ahmad, S., Rukh, G., Varga, T. V., Ali, A., Kurbasic, A., Shungin, D., et al. (2013). Gene × physical activity interactions in obesity: combined analysis of 111,421 individuals of European ancestry. *PLOS Genetics*, *9*(7), e1003607.

27. Andreasen, C. H., Stender-Petersen, K. L., Mogensen, M. S., Torekov, S. S., Wegner, L., Andersen, G., et al. (2008). Low physical activity accentuates the effect of the FTO rs9939609 polymorphism on body fat accumulation. *Diabetes*, *57*(1), 95–101.

28. Qi, Q., Li, Y., Chomistek, A. K., Kang, J. H., Curhan, G. C., Pasquale, L. R., et al. (2012). Television watching, leisure time physical activity, and the genetic predisposition in relation to body mass index in women and men. *Circulation*, *126*(15), 1821–1827.

29. Qi, Q., Chu, A. Y., Kang, J. H., Jensen, M. K., Curhan, G. C., Pasquale, L. R., et al. (2012). Sugar-sweetened beverages and genetic risk of obesity. *New England Journal of Medicine*, *367*(15), 1387–1396.

30. Qi, Q., Chu, A. Y., Kang, J. H., Huang, J., Rose, L. M., Jensen, M. K., & Qi, L. (2014). Fried food consumption, genetic risk, and body mass index: Gene-diet interaction analysis in three US cohort studies. *British Medical Journal, 348*, g1610.

31. Diabetes Prevention Program Research Group. (1999). The Diabetes Prevention Program: Design and methods for a clinical trial in the prevention of type 2 diabetes. *Diabetes Care*, *22*(4), 623–634.

32. Diabetes Prevention Program Research Group. (2002). Reduction in the incidence of type 2 diabetes with lifestyle intervention or metformin. *New England Journal of Medicine*, *346*(6), 393–403.

33. Knowler, W. C., Barrett-Connor, E., Fowler, S. E., Hamman, R. F., Lachin, J. M., Walker, E. A., et al. (2002). Reduction in the incidence of type 2 diabetes with lifestyle intervention or metformin. *New England Journal of Medicine*, *346*(6), 393–403.

34. Wing, R. R. (2010). Long-term effects of a lifestyle intervention on weight and cardiovascular risk factors in individuals with type 2 diabetes mellitus: Four-year results of the Look AHEAD trial. *Archives of Internal Medicine*, *170*(17), 1566–1575.

35. Bray, G., Gregg, E., Haffner, S., Pi-Sunyer, X. F., Wagenknecht, L. E., Walkup, M., & Wing, R. (2006). Baseline characteristics of the randomised cohort from the Look AHEAD (Action for Health in Diabetes) study. *Diabetes & Vascular Disease Research*, *3*(3), 202–215.

36. Pi-Sunyer, X., Blackburn, G., Brancati, F. L., Bray, G. A., Bright, R., Clark, J. M., et al. (2007). Reduction in weight and cardiovascular disease risk factors in individuals with type 2 diabetes: One-year results of the look AHEAD trial. *Diabetes Care*, *30*(6), 1374–1383.

37. Wing, R. R., Bolin, P., Brancati, F. L., Bray, G. A., Clark, J. M., Coday, M., et al. (2013). Cardiovascular effects of intensive lifestyle intervention in type 2 diabetes. *New England Journal of Medicine*, *369*(2), 145–154.

38. Looker, H. C., Knowler, W. C., & Hanson, R. L. (2001). Changes in BMI and weight before and after the development of type 2 diabetes. *Diabetes Care*, *24*(11), 1917–1922.

39. Wadden, T. A., West, D. S., Neiberg, R. H., Wing, R. R., Ryan, D. H., Johnson, K. C., et al. (2009). One-year weight losses in the Look AHEAD study: Factors associated with success. *Obesity*, *17*(4), 713–722.

40. Tsai, A. G., Fabricatore, A. N., Wadden, T. A., Higginbotham, A. J., Anderson, A., Foreyt, J., et al. (2014). Readiness redefined: A behavioral task during screening predicted 1-year weight loss in the look AHEAD study. *Obesity*, *22*(4), 1016–1023.

41. Bouchard, C., Tremblay, A., Despres, J. P., Theriault, G., Nadeau, A., Lupien, P. J., et al. (1994). The response to exercise with constant energy intake in identical twins. *Obesity Research*, *2*(5), 400–410.

42. Delahanty, L. M., Pan, Q., Jablonski, K. A., Watson, K. E., McCaffery, J. M., Shuldiner, A., et al. (2012). Genetic predictors of weight loss and weight regain after intensive lifestyle modification, metformin treatment, or standard care in the Diabetes Prevention Program. *Diabetes Care*, *35*(2), 363–366.

43. McCaffery, J. M., Papandonatos, G. D., Huggins, G. S., Peter, I., Erar, B., Kahn, S. E., et al. (2013). Human cardiovascular disease IBC chip-wide association with weight loss and weight regain in the look AHEAD trial. *Human Heredity*, *75*(2–4), 160–174.

44. McCaffery, J. M., Papandonatos, G. D., Huggins, G. S., Peter, I., Kahn, S. E., Knowler, W. C., et al. (2013). FTO predicts weight regain in the Look AHEAD clinical trial. *International Journal of Obesity*, *37*(12), 1545–1552.

45. Centers for Medicare and Medicaid Services: Decision Memo for Intensive Behavioral Therapy for Obesity (2011). (CAG-00423N).

46. Jensen, M. D., Ryan, D. H., Apovian, C. M., Ard, J. D., Comuzzie, A. G., Donato, K. A., et al. (2013). 2013 AHA/ACC/TOS guideline for the management of overweight and obesity in adults: A report of the American College of Cardiology/American Heart Association Task Force on Practice Guidelines and the Obesity Society. *Circulation, 129*(25 Suppl 2), S139–S140.

47. Kendler, K. S., & Baker, J. H. (2007). Genetic influences on measures of the environment: A systematic review. *Psychological Medicine, 37*(5), 615–626.

48. Christakis, N. A., & Fowler, J. H. (2014). Friendship and natural selection. *Proceedings of the National Academy of Sciences, 111*(Suppl 3), 10796–10801.

49. den Hoed, M., Brage, S., Zhao, J. H., Westgate, K., Nessa, A., Ekelund, U., et al. (2013). Heritability of objectively assessed daily physical activity and sedentary behavior. *American Journal of Clinical Nutrition, 98*(5), 1317–1325.

50. van den Berg, L., Henneman, P., Willems van Dijk, K., Delemarre-van de Waal, H. A., Oostra, B. A., van Duijn, C. M., et al. (2013). Heritability of dietary food intake patterns. *Acta Diabetologica, 50*(5), 721–726.

51. McCaffery, J. M., Papandonatos, G. D., Peter, I., Huggins, G. S., Raynor, H. A., Delahanty, L. M., et al. (2012). Obesity susceptibility loci and dietary intake in the Look AHEAD Trial. *American Journal of Clinical Nutrition, 95*(6), 1477–1486.

52. Lindi, V. I., Uusitupa, M. I., Lindstrom, J., Louheranta, A., Eriksson, J. G., Valle, T. T., et al. (2002). Association of the Pro12Ala polymorphism in the PPAR-gamma2 gene with 3-year incidence of type 2 diabetes and body weight change in the Finnish Diabetes Prevention Study. *Diabetes, 51*(8), 2581–2586.

53. Franks, P. W., Jablonski, K. A., Delahanty, L., Hanson, R. L., Kahn, S. E., Altshuler, D., et al. (2007). The Pro12Ala variant at the peroxisome proliferator-activated receptor gamma gene and change in obesity-related traits in the Diabetes Prevention Program. *Diabetologia, 50*(12), 2451–2460.

54. Keating, B. J., Tischfield, S., Murray, S. S., Bhangale, T., Price, T. S., Glessner, J. T., et al. (2008). Concept, design and implementation of a cardiovascular gene-centric 50 k SNP array for large-scale genomic association studies. *PLoS One, 3*(10), e3583.

55. Voight, B. F., Kang, H. M., Ding, J., Palmer, C. D., Sidore, C., Chines, P. S., et al. (2012). The metabochip, a custom genotyping array for genetic studies of metabolic, cardiovascular, and anthropometric traits. *PLOS Genetics, 8*(8), e1002793.

56. McArdle, P. F., & Whitcomb, B. W. (2009). Improper adjustment for baseline in genetic association studies of change in phenotype. *Human Heredity, 67*(3), 176–182.

57. Li, J., & Ji, L. (2005). Adjusting multiple testing in multilocus analyses using the eigenvalues of a correlation matrix. *Heredity*, *95*(3), 221–227.

58. Storey, J. D., & Tibshirani, R. (2003). Statistical significance for genomewide studies. *Proceedings of the National Academy of Sciences of the United States of America*, *100*(16), 9440–9445.

59. Dabney, A., & Storey, J. D. (2012). Q-value estimation for false discovery rate control. R package version 1.30.0, retrieved from http://www.bioconductor.org/packages/release/bioc/html/qvalue.html

60. Stieger, B. (2009). Recent insights into the function and regulation of the bile salt export pump (ABCB11). *Current Opinion in Lipidology*, *20*(3), 176–181.

61. Wang, R., Salem, M., Yousef, I. M., Tuchweber, B., Lam, P., & Childs, S. J., & Ling, V. (2001). Targeted inactivation of sister of P-glycoprotein gene (spgp) in mice results in nonprogressive but persistent intrahepatic cholestasis. *Proceedings of the National Academy of Sciences, 98*(4), 2011–2016.

62. Henkel, A. S., Kavesh, M. H., Kriss, M. S., Dewey, A. M., Rinella, M. E., & Green, R. M. (2011). Hepatic overexpression of abcb11 promotes hypercholesterolemia and obesity in mice. *Gastroenterology, 141*(4), 1404–1411, 1411 e1401–1402.

63. Andreotti, G., Menashe, I., Chen, J., Chang, S. C., Rashid, A., Gao, Y. T., et al. (2009). Genetic determinants of serum lipid levels in Chinese subjects: A population-based study in Shanghai, China. *European Journal of Epidemiology*, *24*(12), 763–774.

64. Kraja, A. T., Vaidya, D., Pankow, J. S., Goodarzi, M. O., Assimes, T. L., Kullo, I. J., et al. (2011). A bivariate genome-wide approach to metabolic syndrome: STAMPEED Consortium. *Diabetes*, *60*(4), 1329–1339.

65. Bouatia-Naji, N., Rocheleau, G., Van Lommel, L., Lemaire, K., Schuit, F., Cavalcanti-Proenca, C., et al. (2008). A polymorphism within the G6PC2 gene is associated with fasting plasma glucose levels. *Science*, *320*(5879), 1085–1088.

66. Gomez-Ambrosi, J., Rodriguez, A., Catalan, V., & Fruhbeck, G. (2008). The bone-adipose axis in obesity and weight loss. *Obesity Surgery*, *18*(9), 1134–1143.

67. Cheverud, J. M., Ehrich, T. H., Hrbek, T., Kenney, J. P., Pletscher, L. S., & Semenkovich, C. F. (2004). Quantitative trait loci for obesity- and diabetes-related traits and their dietary responses to high-fat feeding in LGXSM recombinant inbred mouse strains. *Diabetes*, *53*(12), 3328–3336.

68. Zhao, L. J., Guo, Y. F., Xiong, D. H., Xiao, P., Recker, R. R., & Deng, H. W. (2006). Is a gene important for bone resorption a candidate for obesity? An association and linkage study on the RANK (receptor activator of nuclear factor-kappaB) gene in a large Caucasian sample. *Human Genetics*, *120*(4), 561–570.

69. Divoux, A., & Clement, K. (2011). Architecture and the extracellular matrix: The still unappreciated components of the adipose tissue. *Obesity Reviews, 12*(5), e494–e503.

70. Qi, L., Qi, Q., Prudente, S., Mendonca, C., Andreozzi, F., di Pietro, N., et al. (2013). Association between a genetic variant related to glutamic acid metabolism and coronary heart disease in individuals with type 2 diabetes. *Journal of the American Medical Association, 310*(8), 821–828.

8 Statistical Strategies for Modeling Gene × Environment Interactions in Longitudinal Cohort Studies

Bhramar Mukherjee, Yin-Hsiu Chen, Yi-An Ko, Zihuai He, Seunggeun Lee, Min Zhang, and Sung Kyun Park

Characterizing gene-environment interaction (GEI) related to quantitative traits associated with common complex diseases in longitudinal cohort studies has received considerable attention in the last decade [1]. In this chapter, we provide an overview of current statistical methods for modeling GEI in longitudinal studies and explore some future directions. We illustrate the methods by using data from the Normative Aging Study, introduced shortly. The remainder of the chapter is organized as follows. In the first section, we review some commonly used methods for modeling the association between longitudinal outcomes of interest (e.g., underlying quantitative traits that are risk factors for a myriad of chronic diseases or intermediate phenotypes like blood pressure or body mass index) and a set of exposures (which could be external environmental factors like air-pollution, pesticide, and heavy metals or behavioral exposures like diet, smoking, and physical activity) measured at baseline or at the same time as the outcome. The exposure set can be a mixed set of time-invariant and time-varying variables. As in the classical repeated measurements framework, properly accounting for intrasubject correlation and modeling the mean outcome trajectory over time, conditional on covariates remain the primary focus of any general longitudinal model, and this basic principle holds for a model including interactions.

Moving forward, in the next section we focus on single-marker-based tests for genetic association and GEI in longitudinal studies. Testing one marker at a time across the genome is straightforward but the primary challenge here is the potential high dimensionality of the genetic data. Common methods for multiple-testing adjustment to control overall type I error rate in genome-wide association studies (GWAS) are typically employed to handle this issue. Related to the single-marker marginal gene-disease association testing, we also discuss the specific issue of testing gene-time/ gene-age interaction. We then focus on methods for jointly modeling the effect of genetic and environmental factors on a longitudinally measured trait of interest. There exists a rich literature on statistical methods for testing GEI in case-control studies. However, there are relatively few papers on efficient modeling of GEI in longitudinal studies thus far. We consider a two-way contingency table layout that is typical of a two-factor analysis of variance (ANOVA) model, and we focus on how to effectively reduce the degrees of freedom to enhance the power of an interaction test. Classical methods that enhance power by reducing the degrees of freedom of the resultant test have been recently adapted to the longitudinal setting by this group, and we discuss these methods.

We then discuss recent set-based tests for genetic association, their extension to longitudinal setting, and the future direction for testing GEI in longitudinal studies using set-based methods. We go on to discuss methods to integrate genetic variants, environmental factors, and time/age in a non-parametric fashion in a longitudinal model. The goal here is to characterize time-dependent interactions in a meaningful way. We discuss methods for identifying differential time contribution to interactions using eigen-decomposition-based models. We follow the convention that each section is accompanied with the relevant references. We conclude with a discussion. Supplementary materials including results of the illustrated methods based on the Normative Aging Study (NAS) data (introduced in the following paragraph) are provided in the appendix.

The Normative Aging Study

The NAS is a multidisciplinary longitudinal study of aging initiated by the US Veterans Administration in 1963 [2]. For illustration purposes, we focus on the longitudinally measured outcome, pulse pressure (PP), which is an important risk factor for arterial stiffness. The environmental risk factor under consideration is bone lead levels measured in tibia bone as measured by the K-x-ray-fluorescence (KXRF) technology, a well-established biomarker for cumulative lead exposure [3]. A relationship between iron deficiency and increased lead absorption is known [4], and increased cumulative lead exposure has been shown to be associated with elevated PP [5]. It is plausible to hypothesize that genes in the iron metabolism pathway could potentially modify the effect of lead exposure on PP. The analysis goal of our pathway-driven GEI study was to test GEI between the cumulative lead exposure and iron metabolism genes on PP. Since 1991, data had been collected every three to five years until 2011, with a median follow-up time of twelve years. The dataset consists of 729 participants from a subset of the NAS data who were successfully genotyped for 22 candidate genes related to iron metabolism and who had measurements of cumulative lead concentrations (μg/g, measured at the tibia bone). A total of 105 single nucleotide polymorphisms (SNPs) on these 22 genes were selected for analysis based on minor allele frequency greater than 0.1. About 95 percent of subjects had repeated measurements on blood pressure, and approximately 50 percent of them had at least four measurements during the study period.

Each of the 105 candidate SNPs had three genotype groups (homozygous major allele, heterozygous, and homozygous minor allele). For illustrating single-gene GEI analysis, two major variants (C282Y and H63D) in the hemochromatosis (*HFE*) gene, the common causal variants of hereditary hemochromatosis [6], were considered for analysis. Given that the research interest was to compare three mutually exclusive groups (wild type for both variants, at least one copy of H63D variant, at least one copy of the C282Y variant), subjects with compound heterozygotes were excluded from analysis following Zhang et al. [5]. Consequently, the *HFE* genotypes were classified into three categories for analysis and the sample size for this analysis was N=671. Henceforth, we let G_1 and G_2 denote the indicator variables for H63D and C282Y, respectively. The environmental exposure (tibia bone lead) was measured as a continuous variable, but when illustrating analysis using categorical environmental exposures, we categorized tibia bone lead concentration into three groups: Low (≤ 15 μg/g), Medium (> 15 and ≤ 25 μg/g), and High (>25 μg/g). Table 8.1 displays cell means of PP and number of participants (in parentheses) for each configuration of the *HFE* genotypes and bone lead levels.

Longitudinal Data Analysis: Basic Strategies

In longitudinal data analysis, random-effect models [7, 8] and generalized estimating equations (GEE) [9, 10] are two widely used methods. Laird and Ware [11] proposed the linear mixed model (LMM) with normally distributed continuous outcome where the random effect terms naturally

Table 8.1

Cell means corresponding to pulse pressure and number of participants (in parentheses) for each configuration of the HFE genotypes and bone lead levels in the Normative Aging Study (Ko et al., 2013)

HFE genotype	Tibia lead levels (µg/g)		
	Low: ≤15	Medium: > 15 and ≤ 25	High: > 25
Wild Type	52.94 (161)	56.16 (149)	56.61 (131)
C282Y	51.89 (23)	56.65 (39)	59.10 (23)
H63D	52.58 (54)	57.72 (53)	64.49 (38)

take into account the between-subject heterogeneity. Williams [12], Breslow [13], and Breslow and Clayton [14] extended the LMM to generalized linear-mixed model (GLMM) that is capable of handling longitudinal data with discrete (binary, ordinal, count) outcome. Suppose the data are composed of n subjects with the jth repeated response corresponding to the ith subject be denoted by y_{ij} and let y_i denote the vector of responses ($n_i \times 1$ vectors); fixed-effect covariates are denoted by x_{ij} ($p \times 1$ vectors), assumed to be measured at the same time as the response with the understanding that for baseline covariates x_{ij} may not depend on j; let the jth repeated set of random-effect covariates be denoted by z_{ij} ($q \times 1$ vectors) for subject i. The GLMM models the conditional distribution of y_{ij} given the random effects as well as covariates and a typical GLMM is written as

$$\eta_{ij} = g(\mu_{ij}) = x_{ij}^T \beta + z_{ij}^T b_i, \tag{8.1}$$

where $\mu_{ij} = E[y_{ij} \mid b_i, x_{ij} z_{ij}]$, b_i is the $q \times 1$ vector of random effects which is typically assumed to follow a multivariate normal distribution with mean 0 and covariance matrix D ($q \times q$ matrix), β ($p \times 1$ vector) is the fixed effect parameter, and $g(\cdot)$ is the link function, linking the conditional mean μ_{ij} to the linear predictor η_{ij}. The subject-specific random effects b_i shared by all observations on subject i are used to induce within-subject correlations.

Some variations such as nonlinear mixed effect models proposed by Lindstrom and Bates [15], skew-normal linear mixed model proposed by Arellano-Valle et al. [16], and log-gamma linear mixed model proposed by Zhang et al. [17] have been developed to accommodate departures from the standard assumptions based on the need for specific applications. Despite the flexibility of the random-effect models, the interpretation of the fixed-effect parameters is subject-specific in a GLMM, and the correspondence with population-based interpretation is only valid for the special case of LMM with identity link function and its modifications with nonnormally distributed random terms.

GEE is an extension of using the marginal linear fixed-effect model with specifications of within-subject correlations to a generalized linear model (GLM) framework. Unlike the multivariate normal distribution, most multivariate distributions cannot be fully specified by the first two moments. Liang and Zeger [18] and Zeger and Liang [19] generalized the quasi-likelihood approach for GLM introduced by Wedderburn [20] to multivariate modeling for nonnormal outcomes. The idea is to rely on unbiased estimating equations (EE) and the asymptotic normality of M-estimators. In order to obtain consistent estimates, only the first moment needs to be correctly specified even with misspecification of the second moment. The lesser the discrepancy between the working correlation and the true correlation structure, the more efficient the estimates are. With similar notations as in (8.1), the general forms of the first moment and second moment as considered in a GEE framework are

$$g\left(\mu_{ij}\right) = x_{ij}^T \beta , \tag{8.2}$$

$$V_i\left(\alpha\right) = \phi A_i^{\frac{1}{2}} R_i\left(\alpha\right) A_i^{\frac{1}{2}} , \tag{8.3}$$

where $\mu_{ij} = E[y_{ij} \mid x_{ij}]$, V_i is an $n_i \times n_i$ covariance matrix, A_i is an $n_i \times n_i$ diagonal matrix with the variance of y_{ij} as the jth element on the diagonal, R_i is an $n_i \times n_i$ "working" correlation matrix, α characterizes the within-subject correlation, and ϕ is the known or unknown scale parameter, which usually is introduced to allow for overdispersion (> 1) or underdispersion (< 1). Note that μ_{ij} in our notation represents the conditional expectation of y_{ij} given random effects b_i and covariates x_{ij} and z_{ij} in GLMM but it refers to the marginal expectation of y_{ij} given covariates x_{ij} in GEE type of methods.

Considerable amount of research has been done to improve the efficiency of the original GEE. Prentice [21] proposed to estimate the parameters of the mean model and the parameters of the covariance matrix concurrently by jointly solving the linear estimating equation for β and quadratic estimating equation for α under GEE framework. It has been shown that the method is more efficient than the GEE first proposed by Liang and Zeger [18] and Zeger and Liang [19]. Qu et al. [22] introduced quadratic inference functions (QIF) motivated from the fact that the inverse of the working correlation can be expressed as a linear combination of some basis matrices. It has been shown that QIF is more efficient than original GEE when the working correlation is misspecified and approximately equally efficient when the working correlation is correctly specified. The advantage/limitation of GEE type of methods is that the inference is population-based, and it is not the approach of choice when individual trajectories and forecasting are of the main interest.

In some applications of longitudinal data analysis, parametric specification of the mean model is not adequate. For example, some biological measurements exhibit nonlinearity as a function of age. Hastie and Tibshirani [23] proposed generalized additive models (GAM) to extend GLM by substituting the linear predictor with additive sums of smooth functions. Lin and Zhang [24] proposed generalized additive mixed model (GAMM) by further extending GAM to accounting for correlated data where the fixed effect terms are modeled nonparametrically. A typical GAMM can be written as

$$g\left(\mu_{ij}\right) = f_1\left(x_{ij1}\right) + f_2\left(x_{ij2}\right) + \ldots + f_p\left(x_{ijp}\right) + z_{ij}b_i \,, \tag{8.4}$$

where $\mu_{ij} = E[y_{ij} \mid b_i, x_{ij}, z_{ij}]$, x_{ijk} is the jth repeated measure of kth covariate for subject i, and $f_k\left(\cdot\right)$ is the centered twice-differentiable smooth function for $k = 1, \ldots, p$.

One distinction important to note here is that the nonlinearity of GAMM refers to the nonlinear relationship between the outcome and the covariates, whereas the nonlinearity in nonlinear mixed effect model refers to the nonlinearity as a function of model parameters. The flexibility of GAMM rises not only from the functional dependence of the outcome on the covariates of interest but also from the choice of smoothers f_k. A semiparametric model is a natural integration of GLMM and GAMM with some fixed effects modeled parametrically while others are modeled nonparametrically. Nonetheless, overfitting can be a problem for typical maximum likelihood estimation in the absence of some restrictions on the smooth term(s). Good and Gaskins [25] proposed maximizing the penalized likelihood with a roughness penalty term in addition to the maximum likelihood function. The smoothing parameter, adjusted to control the smoothness of the underlying function, can be estimated by generalization of the maximum likelihood (GML) given by Anderssen and Bloomfield [26] or more commonly by generalized cross validation (GCV) introduced by Wahba [27]. Fu [28] considered penalized estimating equations, adding a L^γ norm term $\sum_j |\beta_j|^\gamma$ to the estimating equation and shrinking the coefficient estimates toward 0. Penalized GEE have been demonstrated to achieve better efficiency in presence of many correlated covariates.

Presence of certain time-varying covariate is common in a multitude of longitudinal studies and often the effect of the covariate on the outcome of interest is believed to change over time as well. A more generalized class of regression models, varying coefficient model (VCM) introduced by Hastie and Tibshirani [29], in which the coefficients are allowed to change over time as smooth functions is gaining more attention in recent applications [30]. The general form of VCM is:

$$y_{ij} = g\left(\mu_{ij}\right) = \beta_0 + x_{ij1}\beta_1\left(r_{ij}\right) + \ldots + x_{ijp}\beta_p\left(r_{ij}\right), \tag{8.5}$$

where $\mu_{ij} = E[y_{ij} \mid x_{ij}]$, r_{ij} changes the coefficients of $x_{ij1}, x_{ij2}, \ldots, x_{ijp}$ through real-valued functions $\beta_1(\cdot), \ldots, \beta_p(\cdot)$. Variables in r_{ij} may or may not be distinct from the variables $x_{ij1}, x_{ij2}, \ldots, x_{ijp}$. In a many applications, r_{ij} is a time-related variable such as "time" or "age" and time-dependent relationship between longitudinal predictors and longitudinal outcomes is modeled through functions instead of scalars. GLM, in which the coefficients are constant functions of time, is a special case of VCM. VCM can be reduced to GAM with interactions if covariates x_{ij} are categorical.

VCM was further generalized to varying index coefficient model (VICM) by Ma and Song [31]. The linear form of VICM is written as

$$y_{ij} = \beta_0 + x_{ij1}\beta_1\left(a_{ij}^T\gamma_1\right) + \ldots + x_{ijp}\beta_p\left(a_{ij}^T\gamma_p\right) + \epsilon_{ij}, \tag{8.6}$$

where a_{ij} ($s \times 1$ vectors) is the jth repeated set of contributional predictors for subject i and γ_k is a $s \times 1$ vector containing the loading weights for the components of a_{ij}. VICM models the effect of a_{ij} through linear combination of its component. Furthermore, different weights (γ_j) can be assigned to different coefficient functions (β_j). Single-index coefficient model (SICM) proposed by Xia and Li [32] is a special case of VICM by setting common weights to the coefficient-changing variables for different coefficients. In other words, SICM reduces the index coefficient vectors γ_j to a common coefficient vector γ. If $\beta_j(.)$ is a linear function, VICM reduces to a linear regression model with interaction terms but SICM does not. With this general backdrop for longitudinal data, we turn our attention to the specific context of genetic association and gene x environment studies with longitudinal data.

Single-Marker-Based Tests for Genetic Association and Gene-Environment Interaction in Longitudinal Studies

Genome-Wide Association Studies with a Longitudinal Design

Genome-wide association studies (GWAS) have successfully identified many single nucleotide polymorphisms (SNPs) associated with complex traits [33]. However, the fraction of heritability explained by the identified SNPs is low for many complex traits. The large number of SNPs and tests that need to be carried out requires that chosen statistical methods should be computationally feasible and scalable. The typical way for association testing is to test one SNP at a time by a simple linear regression and repeat the same procedure across the genome after correcting for multiple testing. Bonferroni correction can lead to overly conservative thresholds despite its simplicity because of the correlation among the SNPs. Permutation-based methods such as PRESTO [34] and principal component analysis (PCA) related approaches such as Cheverud's M_{eff} [35], Nyholt's M_{eff} [36], and Gao's SimpleM [37], which rely on the effective number of independent tests, were suggested, but none of them has been proven to be uniformly superior across all scenarios, and many are computationally challenging to implement with sparse data. The concerns with multiple testing correction remain identical for longitudinal studies.

When the trait of interest is measured longitudinally, a natural approach for association mapping is to interrogate each SNP individually. We denote the number of alleles corresponding to subject i for SNP g as $G_{ig} = 0, 1,$ or 2 and let y_{ij} be the jth repeated outcome corresponding to subject i. The generic form of the longitudinal model for genetic association can be written as

$$y_{ij} = \beta_0 + \beta_g G_{ig} + x_{ij}^T \beta + \epsilon_{ij}$$

where β_g is the coefficient of genetic effect for SNP g where g can take values from 1 to the number of SNPs considered in the analysis (of the order of millions for a typical GWAS). The error distribution for ϵ_{ij} can be defined appropriately to handle within subject correlation. A linear mixed model as summarized by Furlotte et al. [38] not only accounts for the multiple measurements but also facilitates the investigation of the gene-time/gene-age interaction. Since the genotype of each individual is typically coded as 0, 1, or 2, a separate curve of the trait over time/age for each genotype can be depicted and the interaction can be examined. GEE with robust variance can often be a better choice for running longitudinal GWAS to protect against model misspecification. This has been observed for cross-sectional data [39]. To illustrate the use of longitudinal model for genetic association, we apply the following three models to NAS data.

$$Y_{ij} = \beta_0 + b_{0i} + \beta_{G1}G_{1i} + \beta_{G2}G_{2i} + \beta_t t_{ij} + \beta_{t2}t_{ij}^2 + \beta_C C_i + \epsilon_{ij}$$

$$Y_{ij} = \beta_0 + b_{0i} + \beta_{G1}G_{1i} + \beta_{G2}G_{2i} + \beta_t t_{ij} + \beta_{G1t}G_{1i}t_{ij} + \beta_{G2t}G_{2i}t_{ij} + \beta_C C_i + \epsilon_{ij}$$

$$Y_{ij} = \beta_0 + b_{0i} + f_0(t_{ij}) + G_{1i}f_1(t_{ij}) + G_{2i}f_2(t_{ij}) + \beta_C C_i + \epsilon_{ij}$$

b_{0i} is the random intercept distributed as a normal random variate, G_{1i} and G_{2i} are indicator variables of *H63D* and *C282Y* variant for subject i, t_{ij} is the time elapsed between the 1st visit and the jth visit for subject i, C_i is the age at baseline for subject i, $f_0(\cdot)$, $f_1(\cdot)$, $f_2(.)$ are smooth functions, and ϵ_{ij} is the error term for the measurement corresponding to subject i at the jth visit. We set the errors corresponding to the same subject to follow a multivariate normal distribution with autoregressive order 1 (AR1) covariance structure to flexibly account for intrasubject correlation. The AR1 covariance structure is selected according to Bayesian Information Criterion (BIC). In the first model with quadratic time trend, a negative coefficient for the quadratic term $\widehat{\beta_{2t}}$ (–0.027, p-value < 0.001) suggests that the time effect increases then decreases over time, controlling for the genetic effect. In the second model exploring gene x time interactions on the linear scale for time, $\widehat{\beta_{G2t}}$ (–0.255, p-value = 0.017) is significantly different from 0 at level 0.05 but $\widehat{\beta_{G1t}}$ (–0.011, p-value = 0.884) is not, suggesting that the *H63D* effect is constant over time and the *C282Y* effect linearly decreases over time. The third model, implicitly accounting for nonlinear interactions between genetic factors and time, allows a separate trajectory over time for each of the three genetic groups. The estimated degrees of freedom (EDF) of the smooth curve for the wild-type group ($\widehat{f_0}$) is 3.15, suggesting that the main effect of time is nonlinear, while the EDFs of $\widehat{f_1}$ and $\widehat{f_2}$ are both 2.00, signifying that the gene × time interactions can be adequately represented by cross-product terms between genetic factors and time on a linear scale and more complex nonlinear gene × time model is not needed.

We also applied generalized estimating equation (GEE) approach to NAS data for the first two models and the results are similar in terms of the magnitude of association and statistical significance. The detailed results from the random effects models and those from the GEE approach are available in the supplementary appendix.

Levy et al. [40] proposed to collapse the repeated measurements into a single value by taking arithmetic average and conducted classical association testing. Collapsing longitudinal measurements is not ideal as potential temporal effects and time-dependent interactions cannot be detected. Moreover, with missing observations per subject, a naïve user is often taking average of unequal number of observations per subject, thus creating unequal or nonconstant variance, creating clear misspecification of the standard linear regression model assuming homoscedasticity. Additionally, averaging time-varying exposure and covariate data can lead to loss of power by reducing the variance in exposure data as well as lead to ecological bias.

There is a substantial volume of literature on characterizing gene × time or gene × age interaction. Gong and Zou [41] proposed to investigate gene-time/gene-age interaction by VCM. The genetic effects over time are modeled in a nonparametric fashion and testing for the existence of gene-time/gene-age interaction is the same as testing the constancy of the genetic effects over time. Likelihood ratio (LR) tests based on restricted maximum likelihood (REML) can be conveniently performed to assess whether a time-varying genetic effect or a time-fixed genetic effect is more appropriate.

De Andrade et al. [42] presented a method, extended from the variance components (VC) approach proposed by Hopper and Mathews [43] for evaluating genetic effects and estimating heritability based on the analysis of longitudinal data. In addition, the extended VC approach can accommodate longitudinal familial data. Denote Y_{ikt} as the trait value for k_i members of the ith family and let $E(Y_i) = \mu + X_i\beta$ where $Y_i = (Y_{i1}, Y_{i2}, ..., Y_{iT})^T$ and $Y_{it} = (Y_{i1t}, ..., Y_{ikt})$. The variance component of Y_i is modeled as:

$$V_i = A \otimes G_i + B \otimes \Pi_i + C \otimes I_i, \tag{8.7}$$

where $\otimes$ defines the Kronecker product of two matrices; A, B, and C are the variance-covariance matrix with dimension $T \times T$ for polygenic, the variance-covariance matrix for genes; and the variance-covariance matrix for the covariates of interest, respectively, and G_i, Π_i, and I_i are matrices for pairwise coefficients of relationships between relatives within a family. Testing whether there is gene-time/gene-age interaction is equivalent to testing whether the components of B is equal to 0 or not. One drawback of this method is that the data have to be balanced and the number of variance-covariance terms grow rapidly with T. Wang et al. [44] proposed a semiparametric model for longitudinal traits based on the idea of reduced rank smoothing. The model can deal with the temporal trend of genetic effects and essentially it is a generalization of the VCM for modeling the longitudinal genetic data. Other time-varying covariates besides time/age can be incorporated, and testing for gene-time/gene-age interaction is the same as testing whether the genetic effect is constant or not, similar to the representation in a standard VCM.

Malzahn et al. [45] introduced a longitudinal nonparametric test (LNPT) based on phenotype ranks to test for gene-gene and gene-time interactions, which can be useful for phenotypic traits with nonnormal distributions. The LNPT adjusts for covariates by using categorical variables as whole plot factors. If covariates are correlated, additional procedures (e.g., independent component analysis, using a parametric model for covariate adjustment and then analyzing the residual) are required to construct suitable whole plot factors. The tests for all interactions are performed by applying the respective contrast matrices in an ANOVA-like test statistic, and the corresponding p-values are obtained from central F-distributions. Computations can be carried out by any statistical software package that can calculate midranks of all phenotypes and analyze heteroscedastic factorial designs with repeated measures and unspecified covariance matrices. Given that time is modeled as a categorical variable using this method, it is necessary to find an optimal balance between the number of age categories or follow-up visits and the subgroup sample sizes to evaluate gene-time interaction appropriately.

Longitudinal Models for Gene-Environment Interaction
Typical analysis of repeated measures data attempts to model GEI effects by fitting a regression model to the conditional mean structure of the outcome Y with main effects of genetic factors (G), environmental exposures (E), and G × E interaction terms along with adjustment for other confounders [46, 47]. The standard longitudinal G × E model can be expressed as

$$y_{ij} = \beta_0 + \beta_g G_i + \beta_e E_{ij} + \beta_{ge} G_i E_{ij} + x_{ij}^T \beta + \epsilon_{ij},$$

where G_i is the genetic factor for subject i, E_{ij} is the jth repeated measure of environmental exposure, β_g is the genetic effect, β_e is the environmental effect, and β_{ge} is the interaction effect. When G and E are treated as binary or ordered categorical variables, GEI are often analyzed in the form of a two-way table. Considering G as a row variable with R categories and E as a column variable with C categories, the mean structure of a general two-way classification model for analyzing row × column interactions is given by

$$\mu_{rc} = \mu + \beta_r^G + \beta_c^E + \tau_{rc}, r = 1, \dots, R, c = 1, \dots, C, \tag{8.8}$$

where μ_{rc} is the expected (mean) value of a quantitative trait corresponding to the rth row and the cth column, μ is the grand mean, β_r^G is the additive main effect of the rth row, β_c^E is the additive main effect of the cth column, and τ_{rc} is the nonadditive effect of the rth row and the cth column. The sum-to-zero conditions

$$\sum_r \beta_r^G = \sum_c \beta_c^E = \sum_r \tau_{rc} = \sum_c \tau_{rc} = 0 \tag{8.9}$$

ensure identifiability of the parameters in (8), so the degrees of freedom for testing τ_{rc} in a fully saturated model is $(R-1)(C-1)$. Estimation bias is minimized, since a saturated interaction model in (8.8) does not impose any structural assumptions on the interaction term. However, the number of parameters and hence the corresponding degrees of freedom for the interaction test can increase considerably for finely cross-classified tables. In addition, under a saturated interaction model only observations in a cell can contribute to the parameter estimation for that cell. This may result in reduced efficiency and loss of power for detecting interactions because of small cell sample size in human studies involving a gene with a modest minor allele frequency.

To improve the test for GEI in longitudinal cohort studies, GEI has been explored in alternative parsimonious interaction structures [48, 49] that were originally proposed in the classical analysis of variance (ANOVA) literature. These models were conceived from a statistical objective of reducing degrees of freedom and enhancing power of tests for interaction instead of a mechanistic or biological perspective. We review several classical ANOVA models that have been applied to GEI analysis with the goal of dimension reduction in a two-way GEI table context.

Classical ANOVA Models for GEI

Several classical ANOVA models for testing interaction, originally proposed for data with only one observation per cell, are summarized in the following:

Model (a): $\mu_{rc} = \mu + \beta_r^G + \beta_c^E + \theta \beta_r^G \beta_c^E$ [50]

Model (b): $\mu_{rc} = \mu + \beta_r^G + \beta_c^E + \lambda_r \beta_c^E$ [51]

Model (c): $\mu_{rc} = \mu + \beta_r^G + \beta_c^E + \beta_r^G \eta_c$ [51]

Model (d): $\mu_{rc} = \mu + \beta_r^G + \beta_c^E + \theta \beta_r^G \beta_c^E + \lambda_r \beta_c^E + \beta_r^G \eta_c$ [52]

with constraints $\sum_{r=1}^{R} \beta_r^G = \sum_{c=1}^{C} \beta_c^E$ for models (b) and (c), respectively, and additional constraints $\sum_{r=1}^{R} \lambda_r \beta_r^G = \sum_{c=1}^{C} \eta_c \beta_c^E = 0$ for model (d). Among the four models, the multiplicative interaction terms are proportional to the main effects of one or both factors. The null hypotheses of no interaction for models (a)-(d) are $\theta = 0$, $\lambda_r = 0 (r = 1,...,R-1)$, $\eta_c = 0 (c = 1,...,C-1)$, and $\theta = \lambda_r = \eta_c = 0$ $(r = 1,...,R-2; c = 1,...,C-2)$, respectively.

Given that the interaction terms in models (a)-(d) are functions of row and/or column main effects, the existence of interaction is conditional on the presence of main effects. Notice that the interaction is not identifiable when main effects are not present. When the candidate genes to be studied for GEI are selected based on marginal genetic associations [53, 54], it may be reasonable to adopt one of these interaction forms for GEI. Compared to a saturated interaction model for GEI, models (a)–(d) may offer superior efficiency in terms of interaction parameter estimation and test (due to reduced degrees of freedom) if the model is correctly specified. The consequence of fitting inappropriate or misspecified statistical models to GEI, however, is that epidemiologists or geneticists may fail to detect important interaction effects.

Additive Main Effects and Multiplicative Interaction (AMMI) Models

Alternatively, a lower rank representation of the interaction matrix may provide a solution to model misspecification problems of GEI. Gollob [55] proposed a factor-analysis of variance (FANOVA) model to decompose a two-way table. The essential idea is to represent the $R \times C$ interaction matrix Γ with interaction parameters τ_{rc} as entries, by the following representation:

$$\Gamma = ADB^T.$$

Here A and B are $R \times s$ and $C \times s$ orthonormal matrices ($A^T A = B^T B = I$) and D is a $s \times s$ diagonal matrix with elements $d_1 \geq d_2 \geq \ldots \geq d_s$. The maximum rank of Γ is $min(R-1, C-1)$ because of the sum-to-zero constraints on τ_{rc}. This makes the matrix Γ doubly centered. Let $R \leq C$, thus the maximal rank of Γ is $R-1$. By this factor representation, for a saturated interaction model, τ_{rc} is perfectly reproduced by,

$$\tau_{rc} = \sum_{m=1}^{R-1} d_m \alpha_{rm} \gamma_{cm}$$

However, one can think of a sparse representation of the interaction matrix by retaining the first $M < R-1$ components of this representation, namely,

$$\tau_{rc} = \underbrace{\sum_{m=1}^{M} d_m \alpha_{rm} \gamma_{cm}}_{\text{leading term}} + \underbrace{\phi_{rc}}_{\text{random noise}} \qquad (8.10)$$

This representation in (8.10) gives rise to the general class of additive main effects, multiplicative interaction models (AMMI) [56, 57, 58, 59, 60]. AMMI models entail singular value decomposition (SVD) of the cell residual matrix after fitting the additive main effects, which consists of a sum of several successive multiplicative contrasts with each contrast orthogonal to all previous contrasts. The models do not have structural assumptions on the interaction term. By choosing a small number of principal interaction components, one is able to reduce the effective degrees of freedom of the resultant interaction test. An AMMI model is given by

$$\text{Model (e): } \mu_{rc} = \mu + \beta_r^G + \beta_c^E + \sum_{m=1}^{M} d_m \alpha_{rm} \gamma_{cm} + \phi_{rc},$$

where M represents the number of interaction factors being extracted, $M \leq min(R-1, C-1)$, and a residual ϕ_{rc} remains if not all interaction factors are used. Note that α_{rm} and γ_{cm} are not functions of the main effects β_r^G and β_c^E but are independent parameters. The terms $\{\alpha_{rm}\gamma_{cm}\}$ can be considered as the weights corresponding to a multiplicative contrast among $\{\tau_{rc}\}$ with $\sum_r \alpha_{rm} = \sum_c \gamma_{cm} = 0$ and $\sum_r \alpha_{rm}\alpha_{rm'} = \sum_c \gamma_{cm}\gamma_{cm'} = 0$ for $m \neq m'$. For a fixed M and normalized contrasts ($\sum_r \alpha_{rm}^2 = \sum_c \gamma_{cm}^2 = 1$), $\{d_m, \alpha_{rm}, \gamma_{cm}\}$ can be obtained by applying SVD to $\{\tau_{rc}\}$ with d_m the singular value for the mth principal component and α_{rm}, γ_{cm} the corresponding left and right singular vectors, respectively. The null hypothesis of no interaction for AMMI model is $H_0 : d_m = 0, m = 1, \dots, M$.

AMMI has been shown to perform well across a spectrum of interaction structures, either based on naive F test using subject-level mean [49] or a parametric bootstrap approach [48] Thus, AMMI model should be considered as a useful screening tool for detecting interaction effects even in the absence of genetic main effects.

Tests for GEI Using Models (a)–(e)

In considering the application of models (a)–(e) to GEI tests with repeated measures data, Mukherjee et al. [49] proposed a screening tool for interaction using cell means from an unbalanced repeated measures array. This approach is appealing due to a closed-form analytical expression of the test statistic. However, violations in the homoscedasticity assumption of cell mean error distributions result in inflated type 1 error.

Two approaches have been proposed to overcome the limitations of analyzing cell means [48]. One is using LRTs for GEI but still based on summarized cell means, and the other is using parametric bootstrap resampling based on individual observations. In the cell mean based method, an empirical variance estimate for each cell mean is used by assuming a within-subject correlation structure to account for within-subject correlation resulting from repeated measurements. For classical interaction models (a)–(d) involving nonlinearity in the parameters, the MLEs for main and interaction effects can be obtained using a quasi-Newton method [61], which is a sequential line search algorithm and generally requires only the gradient of the objective to be computed at each iteration. When convergence is reached, the log-likelihood under the null ($\hat{l}_0$) and under the alternative ($\hat{l}_1$) are used to construct the LRT statistic: $-2(\hat{l}_0 - \hat{l}_1)$. Under H_0, the LRT statistic approximately follows a central chi-square distribution with degrees of freedom 1, $R-1$, $C-1$, and $R+C-3$ for models (a), (b), (c), and (d), respectively.

Regarding the parameter estimation for model (e), a direct SVD solution does not exist because of unequal variances due to unbalanced repeated measures data. Instead, MLEs of the interaction matrix can be obtained using crisscross regression [62] by transforming the estimation problem to an iterative weighted least squares problem. Once the MLEs are obtained, the LRT proposed by Boik [63] for testing the rank of interaction matrix (unbalanced data without repeated measures) can be modified for application to GEI using AMMI models [48]. The asymptotic null distribution of this LRT statistic converges in distribution to the maximum root of a Wishart matrix with degrees of freedom $max(R-1, C-1)$ in balanced designs [63]. Under the assumption of proportional balanced data, the null distribution of the LRT for unbalanced data is known to be identical to that in balanced designs. However, due to correlated nature of the outcome data, this null distribution is not applicable to repeated measures data.

In the presence of confounders and other covariates, a mixed-effects regression model using all individual observations provides a general framework for handling repeated measurements. To avoid computationally intensive iterations associated with estimation for models (a)-(e), one can use a noniterative two-step regression procedure to approximate the interaction parameter estimates [48]. For classical interaction models (a)–(d) where the interactions are a function of main effects, the main effects are removed in the initial step by fitting a saturated interaction model. Instead of estimating main effects by fitting a null-hypothesis (no interaction) model [64], the main effect estimates are obtained from a saturated model. As such, any existing interaction effect would not bias the estimation of main effect parameters in unbalanced data settings when the parameter estimates are not orthogonal. In the second step, the residuals are regressed on specific forms of main effect estimates (without intercept) to obtain the corresponding slope estimates for the interaction effects. Any error distribution is assumed to be preserved in the second step. Given that the interaction structure of model (e) is derived from a SVD of the matrix of residuals after removing additive effects, a SVD is performed to the estimated saturated interaction matrix $\hat{\Gamma}$. The resulting largest singular value of $\hat{\Gamma}$ is an approximation of $\hat{d}_1$. The corresponding left and right singular vectors are approximations of $\hat{\alpha}_r$ and $\hat{\gamma}_c$, for $r = 1, ..., R$, $c = 1, ..., C$. Subsequently, one may construct a LRT-based pivot using these noniterative two-step regression estimates for models (a)–(e). Since the LRT-based pivot does not have a standard asymptotic distribution, parametric bootstrap is used to elicit the null distribution of this LRT-based pivot [48].

Shrinkage Estimator for GEI Using Tukey's and Saturated Models

As previously mentioned, a saturated interaction model in (8.8) consists of parameter estimation for each configuration of GEI. Though unbiased, the estimates are inefficient, especially for small samples of certain GEI configurations in an observational GEI study. Compared to a saturated interaction model, model (a) in particular can lead to increased efficiency since all observations are used to estimate the interaction parameter. However, when the assumption of the interaction form in model (a) is violated (e.g., absence of genetic main effects), the estimate for the interaction effect will be biased and the corresponding one degree-of-freedom test can result in extremely low power [49, 66]. To achieve a counterbalance between bias and efficiency in detecting GEI in longitudinal cohort studies, one can model the interaction structure with a shrinkage estimator using estimates from model (a) and a saturated interaction model [65].

Denote the interaction parameters to be estimated for a $R \times C$ GEI table by

$$\tau = (\tau_{11}, \tau_{21}, \ldots, \tau_{(R-1)1}, \tau_{12}, \ldots, \tau_{(R-1)(C-1)})^T$$

Let τ_{tuk} and τ_{sat} be the asymptotic limits of the estimator of τ from model (a) and saturated interaction model in (8.8), respectively, each being a length $(R-1)(C-1)$ vector. When the true model is model (a), we have $\tau_{tuk} - \tau_{sat} = \delta = 0$. To relax the model assumption, let $\delta \sim N(0, \Theta)$. A conservative estimate of Θ is given by $\hat{\delta}\hat{\delta}^T$, where $\hat{\delta} = \hat{\tau}_{tuk} - \hat{\tau}_{sat}$ and $\hat{\tau}_{tuk} = \hat{\theta}(\hat{\beta}_1^G \hat{\beta}_1^E, \hat{\beta}_2^G \hat{\beta}_1^E, \ldots, \hat{\beta}_{R-1}^G \hat{\beta}_{C-1}^E)^T$. Let the shrinkage factor be $B = \hat{V}_\tau (\hat{V}_\tau + \hat{\delta}\hat{\delta}^T)^{-1}$, where $\hat{V}_\tau$ is the estimated variance-covariance matrix of $\hat{\tau}_{sat}$. Then the shrinkage estimator for the interaction effect τ is given by

$$\hat{\tau}_{shk} = \hat{\tau}_{sat} + B(\hat{\tau}_{tuk} - \hat{\tau}_{sat}), \tag{8.11}$$

where $\hat{\tau}_{tuk}$ and $\hat{\tau}_{sat}$ are MLEs from Tukey's model (a) and the saturated interaction model, respectively.

The shrinkage factor B in (11) determines the amount of shrinkage of $\hat{\tau}_{sat}$ toward $\hat{\tau}_{tuk}$. As $\hat{\delta} \to 0$ and $\hat{B} \to I$, $\hat{\tau}_{shk} \to \hat{\tau}_{tuk}$ (data are indicative of a Tukey's interaction structure). On the other hand, as the bias of Tukey's model estimator $\hat{\delta}$ increases, the largest eigenvalue of $\hat{B}$ goes to 0 and $\hat{\tau}_{shk} \to \hat{\tau}_{sat}$ (data are not in favor of Tukey's form of interaction). Now express the shrinkage estimator in (8.11) as

$$\hat{\tau}_{shk} = \hat{\tau}_{sat} + \hat{V}_\tau \left(\hat{V}_\tau^{-1} - \frac{\hat{V}_\tau^{-1}\hat{\delta}\hat{\delta}^T\hat{V}_\tau^{-1}}{1 + \hat{\delta}^T\hat{V}_\tau^{-1}\hat{\delta}} \right) \hat{\delta} = \hat{\tau}_{sat} + \hat{\delta} - \hat{\delta} \left(\frac{\hat{\delta}^T\hat{V}_\tau^{-1}\hat{\delta}}{1 + \hat{\delta}^T\hat{V}_\tau^{-1}\hat{\delta}} \right)$$

When data are under Tukey's model, $\hat{\delta} \to 0$ as $N \to \infty$. When data are not under Tukey's model, the largest eigenvalue of $\hat{V}_\tau$ goes to 0 and $\hat{\delta}^T \hat{V}_\tau^{-1} \hat{\delta} \to \infty$ as $N \to \infty$. So, the term $\left(\hat{\delta}^T \hat{V}_\tau^{-1} \hat{\delta}\right)/(1 + \hat{\delta}^T \hat{V}_\tau^{-1} \hat{\delta})$ converges to 1. This indicates that $\hat{\tau}_{shk}$ is asymptotically equivalent to $\hat{\tau}_{sat}$, which is an unbiased estimator of τ. But with moderate sample size, $\hat{\delta}$ creates a small bias in $\hat{\tau}_{shk}$ that can be traded for a larger decrease in variance, leading to an improvement in finite sample mean squared error [67]. In addition, when main effects are not present, the shrinkage estimator will guard against the instability of parameter estimates under model (a) by shrinking $\hat{\tau}_{shk}$ toward $\hat{\tau}_{sat}$.

Given an empirical variance estimate of the shrinkage estimator, the Wald test may be used as an approximate test for interaction. The test statistic for $H_0 : \tau = 0$ is given by $\tilde{T}_W = \hat{\tau}_{shk}^T \hat{\Sigma}_{\hat{\tau}_{shk}}^{-1} \hat{\tau}_{shk}$, where $\hat{\Sigma}_{\hat{\tau}_{shk}}$ is the estimated variance-covariance matrix for $\hat{\tau}_{shk}$. $\hat{\Sigma}_{\hat{\tau}_{shk}}$ can be approximated by multivariate Taylor series expansion (or the delta method). $\tilde{T}_W$ approximately follows a χ^2 with degrees of freedom $(R-1)(C-1)$ under H_0.

Data Analysis

First, we apply models (a)–(e) to investigate the interaction effect between the polymorphisms of the *HFE* gene and cumulative lead exposure using the methods proposed in Ko et al. [48]. Then, we illustrate the use of the shrinkage estimator discussed earlier to test GEI effects between 105 SNPs on genes in the iron metabolic pathway and cumulative lead exposure. The results will be compared with a saturated interaction model and Tukey's model.

LRT-CM and LRT-PB [48]

Let y_{rck} denote the length-n_{rck} observation vector for the kth subject in the (r,c)th cell of a $R \times C$ GEI table. According to the Akaike information criterion (AIC) for model fit, a random-intercept mixed-effects model was chosen for analysis:

$$y_{rck} = \mu_{rc} 1_{n_{rck}} + b_{rck} 1_{n_{rck}} + e_{rck}, \tag{8.12}$$

where μ_{rc} is the mean response value for the (r, c)th cell, $\mu_{rc} = \mu + \beta_r^G + \beta_c^E + \tau_{rc}$, $1_{n_{rck}}$ is a vector of ones with length n_{rck}, $b_{rck} \sim N(0, \sigma_b^2)$ is the random-effect coefficient for subject (r, c, k), $e_{rck} \sim N(0, \sigma_e^2 I)$ is the random error term, and $\{\sigma_b^2, \sigma_e^2\}$ are assumed to be independent and constant across individuals. The model adjusting for baseline age, time since baseline in years, and squared time was used:

$$y_{rck} = \mu_{rc}1_{n_{rck}} + \beta_{Age}Age_{rck} + \beta_{Time}Time_{rck} + \beta_{Time^2}Time_{rck}^2 + b_{rck}1_{n_{rck}} + e_{rck} \quad (8.13)$$

For LRT-CM adjusting for covariates, cell means were formed by the residuals from a regression of the outcome on covariates other than G and E. This is an ad-hoc approach for covariate adjustment, since correlations of covariates with G and E are ignored. In general, LRT-PB based on a full regression model with G, E, and covariates will yield more power.

Table 8.2 shows the p-values for testing $HFE \times$ lead exposure interaction using LRT-CM and LRT-PB and the saturated interaction model. Because the maximum rank of the interaction matrix is two, here model (e) refers to AMMI model with $M = 1$. Using LRT-CM without adjustment for any covariate, the interaction was significant in all four classical models (p – value < 0.05), whereas (e) gave a p-value between 0.05 and 0.10. After adjusting for the covariates, model (e) detected the interaction using LRT-CM (p – value < 0.01). Regardless of covariate adjustment, the interaction was significant for models (a)–(e) using LRT-PB (p – value < 0.01), and

Table 8.2
Analysis results of GEI between HFE genotypes and tibia lead levels in the Normative Aging Study using the likelihood ratio test with cell means (LRT-CM) and the parametric bootstrap (LRT-PB) approach. 1,000 replicates were simulated under the null hypothesis (Ko et al., 2013)

| Model | Hypothesis | p-value | | | |
		LRT-CM[a]	LRT-CM[b]	LRT-PB[a]	LRT-PB[b]
Model (a)	$H_0 : \theta = 0$	0.008	0.002	0.002	0.003
Model (b)	$H_0 : \lambda_i = 0$ (Lead)	0.029	0.007	0.009	0.008
Model (c)	$H_0 : \eta_i = 0$ (HFE)	0.015	0.002	0.002	0.001
Model (d)	$H_0 : \theta = \lambda_i = \eta_i = 0$	0.035	0.005	0.007	0.002
Model (e)	$H_0 : d_1 = 0$	<0.10	<0.01	0.009	0.002
Saturated Model	$H_0 : \tau_{ij} = 0$			0.015	0.006

a. No covariate adjustment.
b. Adjusting for baseline age, time since baseline, and squared time. For LRT-CM, residuals from a regression of pulse pressure on all other covariates except lead levels and genotype were used to form the cell means.

for the saturated interaction model (p – value < 0.02). P-values for the GEI effect decreased further for all tests with covariate adjustment. Given the significant GEI on all models, this interaction may be real and not model dependent.

Shrinkage Estimator

Now we revisited the study to include 105 SNPs in 22 genes in the iron metabolic pathway to test for GEI using the shrinkage estimation framework [65]. We used Tukey's model, saturated interaction model, and the shrinkage approach to model the GEI structures for each SNP × lead interaction with model (13). Given that these SNPs are located in a small number of genomic regions, they are in close proximity to each other and thus may exhibit linkage disequilibrium (LD). To control for the family-wise error rate while accounting for the potentially correlated SNPs in the multiple testing procedure, we adjusted the significance level according to the effective number of independent tests (denoted by M_{eff}) using the simple M method [37]. This method involves first estimating the correlation matrix among the 105 SNPs by the composite LD, calculating the corresponding eigenvalues, and then finding M_{eff} through principal component analysis. We chose M_{eff} = 89 so that the corresponding eigenvalues explained at least 99.5 percent of the variation for the SNP data. Thus, the adjusted significance level was $0.05/M_{eff}$ = 0.05/89 = 5.6×10^{-4}. Table 8.3 lists the smallest p-values of GEI tests for the three top-ranked SNPs by using Tukey's model, the shrinkage estimator, and saturated interaction model within iron gene regions in the NAS data. The Wald test via the shrinkage estimator yielded

Table 8.3

The p-values of gene × environment interaction tests for the top three (ranks in parentheses) single-nucleotide polymorphisms by using Tukey's model, the shrinkage estimator, and the saturated interaction model within iron gene regions in the Normative Aging Study data (adjusted α = 5.6×10^{-4}) (Ko et al., 2014)

SNP ID	Gene	Tukey LRT	Shrinkage Wald	Saturated LRT
rs1799945	SEC16B	0.003 (1)	1×10^{-4} (1)	0.006 (1)
rs2285228	DMT1	0.005 (2)	0.001 (2)	0.017 (2)
rs3821716	MFI2	0.014 (3)	0.012	0.120
rs422982	DMT1	0.016	0.003 (3)	0.072 (3)

the smallest p-values across all top ranked SNPs listed in the table (and one reached statistical significance), compared with Tukey's and saturated interaction models. We found a significant modifying effect of SNPrs1799945 in the *HFE* gene using the shrinkage estimator ($p = 10^{-4}$) that replicates our previous findings. For the wild-type participants, mean PP remained nearly unchanged between the High and the Low tibia lead groups. In contrast, mean PP was estimated to be 7.35 mmHg (95 percent CI = [6.53, 8.17]) higher for the High tibia lead group than the Low tibia lead group among the homozygous mutant carriers.

Set-Based Tests for Genetic Association and Gene-Environment Interaction in Longitudinal Studies

Set-Based Tests for Genetic Association

Presence of rare variants (RVs) with extremely low frequencies in next generation sequencing studies results in more challenges for identifying new genetic associations to explain additional heritability. To reduce the burden of the adjustment of multiple testing and boost power, an alternative strategy is to jointly model a set of variants (e.g., all variants in a gene), instead of modeling each variant individually, resorting to GLM as

$$\eta_i = g(\mu_i) = \beta_0 + X_{i1}\beta_1 + \ldots + X_{iK}\beta_K \tag{8.14}$$

and the hypothesis of no association is whether all the K coefficients are equal to 0, which involves a global test of K degrees of freedom. The test may be underpowered, especially when K is large and the inclusion of some unassociated RVs in the model. The single marker based test and the global test are at two extremes with two conflicting goals—reducing degrees of freedom and adjustment of multiple testing. The methods summarized in the rest of this section focus on reconciling the two goals by some dimension reduction techniques.

Chapman and Whittaker [68] and Pan [69] proposed Sum-test, which assumes that all the SNPs are associated with the trait with the same magnitude as the following:

$$\eta_i = g(\mu_i) = \beta_0 + \sum_{k=1}^{K} X_{ik}\beta \tag{8.15}$$

The hypothesis testing now is a one-degree-of-freedom test, which potentially leads to higher power. The testing can be easily conducted by likelihood ratio test (LRT), score test, or Wald test. The cohort allelic sums test (CAST) proposed by Morgenthaler and Thilly [70] is closely related to Sum-test with a different coding system for the "super variant" that collapses multiple rare variants in a gene. The weighted sum test (w-Sum) proposed by Madsen and Browning [71] that jointly analyzes a group of variants to test for groupwise association with the trait is an adaptive version of the original Sum-test with minor allele frequency (MAF) based weighting scheme. Nevertheless, when some associations are in opposite directions, the Sum-test suffers from power loss. Han and Pan [72] proposed a heuristic approach to avoid the consequence of the β_k with opposite directions. The idea is to flip the coding of the SNPs with highly negative coefficients obtained from the SNP-by-SNP test.

Pan [69] proposed the following two statistics based on the components of effect estimates $\beta_k s$ from single-variant tests without making extra assumptions such as homogeneous magnitude of association made in Sum-test.

$$SumSqB = \sum_{k=1}^{K} \hat{\beta}_k^2 \tag{8.16}$$

$$SumSqBw = \sum_{k=1}^{K} \frac{\hat{\beta}_k^2}{\hat{v}_k} \tag{8.17}$$

$\hat{v}_k$ is the estimated variance of $\hat{\beta}_k$ from SNP-by-SNP model. The two statistics have a quadratic form of normal variates under the null so the testing can be easily performed accordingly. Pan [69] also proposed two modified score tests analogous to $SumSqB$ and $SumSqBw$ based on (8.14) as the following:

$$SumSqU = U_M^T U_M = (Y - \bar{Y})^T X X^T (Y - \bar{Y}) \tag{8.18}$$

$$SumSqUw = U_M^T Diag[\bar{Y}(1 - \bar{Y})(X - \bar{X})^T (X - \bar{X})]^{-1} U_M \tag{8.19}$$

where $U_{M,k} = X_k^T(Y - \bar{Y})$. The two statistics both have quadratic forms but different from usual scores test statistics because the covariance term is replaced by a diagonal working covariance matrix. Nonetheless, the statistics under the null hypothesis can be approximated by $a\chi_d^2 + b$ where d is the estimated degrees of freedom. The preceding four tests, which are not sensitive to opposite directions of the variants, can be regarded as modifications of Sum-test with higher power in certain scenarios.

Li and Leal [73] proposed a test called combined multivariate and collapsing (CMC) test to unify the two extremes—multiple-marker testing such as Hotelling's T^2 test and collapsing methods such as Sum-test. The notion of CMC approach is to divide the variants into subgroups on the basis of a certain predetermined criteria such as allele frequencies and the variants are collapsed within each subgroup. A multivariate test with G (the number of subgroups) degrees of freedom is then conducted. One drawback of CMC is that the choice of classification criteria is subjective. Furthermore, potential confounders cannot be easily controlled in the analysis.

Liu and Leal [74] proposed kernel-based adaptive clustering (KBAC) for detecting complex trait associations with rare variants. Suppose we have $M + 1$ mutation patterns $(G_0, G_1, ..., G_M)$ across all k variants and assume there are n_{i1} cases and n_{i0} controls with mutation pattern G_i. The KBAC test statistic is

$$T_{KBAC} = \left[\sum_{i=1}^{M} (n_{i1} / n_1 - n_{i0} / n_0) w_i \right]^2,$$ (8.20)

where the weights w_i can be assigned based on one of the three asymptotically equivalent kernels—hypergeometric kernel, marginal binomial kernel, and asymptotic normal kernel. Similar to the Sum-test, the performance of KBAC may be deteriorated with the presence of both protective and harmful genetics variants. A simple remedy that may help overcome the problem is to take the square of the ratio difference before weighting. KBAC has been shown to be attractive in detecting possible interactions among the variants.

Neale et al. [75] proposed C-alpha test to test the association between a binary trait/disease and genetics variants. The test statistic T contrasts the variance of each observed count with the expected variance:

$$T = \sum_{i=1}^{M} [(y_i - n_i p_0)^2 - n_i p_0 (1 - p_0)]^2,$$ (8.21)

in which n_i is the number of times that variant i is observed, y_i is the number of cases out of n_i that variant i is observed, and p_0 is the probability of being a case subject under the null (proportion of cases in the sample). The test statistic converges to a standard normal distribution as M, the number of variants, goes toward infinity under the null hypothesis. P-value can be obtained by permutation of case-control status or normal approximation. C-alpha tests the variance rather than the mean. When the set of SNPs contain both protective and harmful variants, C-alpha test is more sensitive to detect the variants in the same gene.

A kernel-machine regression based method called sequence kernel association test (SKAT) was proposed by Wu et al. [76], and the C-alpha test was shown as a special case of SKAT. The kernel-machine framework has become very popular for modeling high-dimensional genetics data because its flexibility to handle nonlinear relationship between complex traits and variants while controlling for covariates. The general form of a semiparametric SKAT model is given by

$$\eta_i = g(\mu_i) = \beta_0 + \beta_1 Z_{i1} + \ldots + \beta_p Z_{ip} + f(X_{i1}, X_{i2}, \ldots, X_{iK}), \tag{8.22}$$

where Z_{i1}, Z_{i2}, ..., Z_{ip} are the covariates to adjust for and $f(.)$ is an arbitrary function defined by a positive, semidefinite kernel function $K(.,.)$. The popular choices of kernels are linear (weighted) kernel, identity by state (IBS) kernel, and Gaussian kernel. In order to test whether there is an association between the SNP set and the phenotype, we consider the null hypothesis $H_0 : f(X) = 0$. The connection between the kernel-machine framework and GLMM has been exploited by Liu et al. [77] and testing the hypothesis is equivalent to testing the variance component, i.e., $H_0 : \tau = 0$, if we treat f as a subject-specific random effect via the GLMM connection where f follows an arbitrary distribution with mean zero and variance τK. If we rely on the variance-component score test proposed by Zhang and Lin [78], the SKAT statistic becomes

$$Q = \frac{(Y - \hat{p}_0)^T K(Y - \hat{p}_0)}{2}, \tag{8.23}$$

in which $\hat{p}_0$ is the predicted mean under the null hypothesis, with $g(\hat{p}_{0i}) = \hat{\beta}_0 + \hat{\beta}_1 Z_{i1} + \ldots + \hat{\beta}_p Z_{ip}$. The reference distribution of Q under the null hypothesis is a mixture of χ^2 distributions.

The rationale behind SKAT and other set-based tests is that they employ prior knowledge to group multiple SNPs into SNP sets on the basis of their proximity to a known gene and reduce multiple comparisons and harness local linkage disequilibrium (LD) to improve power. Some other test like replication-based test was proposed by Ionita-Laza et al. [79] to deal with opposite association directions. A fairly thorough simulation study for comparing some of the above-mentioned genetics association testing models was conducted by Basu and Pan [80].

From a random field framework and development in spatial statistics, He et al. [81] proposed the genetic random field model (GenRF) for modeling and testing joint association. In this approach, outcomes (phenotypic values) are viewed as stochastic realizations of a random field on a genetic space where the genotype sequences determine the location and the outcomes are correlated depending on their spatial location, e.g., the closer the more similar if the association exists. The GenRF model is

$$E(Y_i|Y_{-i}) = Z_i^T \beta + \gamma \sum_{j \neq i} s_{i,j} (Y_j - Z_j^T \beta), \tag{8.24}$$

where Y_i and Z_i are the phenotype and covariates of subject i; Y_{-i} denotes phenotypes for all other subjects except Y_i; $Z_i^T \beta$ is the covariates contribution; $s_{i,j}$ is a known weight, determined by the genetic similarity between subjects i and j; and γ is a coefficient measuring the magnitude of the overall contribution of Y_{-i}, i.e., the magnitude of the joint association between the K genetic variants and the outcome. To test the association, the null hypothesis is $H_0 : \gamma = 0$ involving a single parameter. The genetic similarity plays the same role as the kernel function in SKAT, but there is no need to be positive and semidefinite. Popular choices include genetic relationship [82] and IBS similarity. By a pseudo-likelihood approach, the GenRF test statistic is

$$\hat{\gamma} = \frac{(Y - \hat{\mu})^T S (Y - \hat{\mu})}{(Y - \hat{\mu})^T S^2 (Y - \hat{\mu})}, \tag{8.25}$$

where $\hat{\mu}$ is the estimated mean under H_0; S is the genetic similarity matrix in which the (i,j)th cell is $s_{i,j}$ when $i \neq j$, and 0 when $i = j$. Intuitively, a large value of $\hat{\gamma}$ would give the evidence to reject H_0. Simulation studies showed that the method has overall comparable performance as SKAT but is less conservative under H_0, and the power difference depends on specific factors, such as presence of interactions, structure of linkage disequilibrium, and percentage of causal variants [83].

Though set-based methods have gained popularity, their extension to longitudinal study is still rare due to the complexity of accounting for the intrasubject correlation and modeling the mean outcome trajectory over time. He et al. [84] extended the genetic random field model to longitudinal setting and proposed a within-subject similarity to model the intrasubject correlation. The longitudinal genetic random field model (LGRF) is

$$Y_{i,u} \mid Y_{-(i,u)} = Z_{i,u}^T \beta + \sum_{v \neq u} w(t_{i,v}, t_{i,u}; \eta)(Y_{i,v} - Z_{i,v}^T \beta) + \gamma \sum_{(j,v) \neq (i,u)} s_{i,j}(Y_{j,v} - Z_{j,v}^T \beta) + \varepsilon_{i,u},$$

$$(8.26)$$

where $Y_{i,u}$ and $Z_{i,u}$ are the outcome and covariates of subject i at measurement u; $Y_{-(i,u)}$ denotes all other phenotypic values except $Y_{i,u}$; $Z_{i,u}^T \beta$ is the covariates contribution to the outcome mean from nongenetic covariates; $\varepsilon_{i,u} \sim i.i.d \, N(0, \sigma^2)$; $s_{i,j}$ is the genetic similarity between subjects i and j as in the genetic random field model; γ measures the magnitude of the genetic association; $w(t_{i,v}, t_{i,u}; \eta)$ is the within-subject similarity between $\tilde{Y}_{i,u}$ and $Y_{i,v}$ with parameters η introducing the intrasubject correlation. For example, if two measurements are measured closer in time, $w(t_{i,v}, t_{i,u}; \eta)$ may be larger. Interestingly, the within-subject similarity specifies a precision matrix (the inverse of a covariance matrix) instead of the covariance matrix in a usual GEE method. They showed that this specification might have insight connection with the quadratic inference function approach [22] where they consider the precision matrix as a linear combination of several basis matrices to approximate the intrasubject correlation structure. To test $H_0 : \gamma = 0$, a generalized score statistic is proposed as

$$Q_G = \frac{(Y - \hat{\mu})S[I - W(\hat{\eta})](Y - \hat{\mu})}{m},$$

$$(8.27)$$

where m is the number of subjects; $\hat{\mu}$ and $\hat{\eta}$ are estimated under H_0; S is the genetic similarity matrix, which consists of $m \times m$ blocks, and all elements of the (i,j)th block are $s_{i,j}$. The diagonal elements of S are 0; $W(\eta)$ is a block diagonal matrix of the within-subject similarity, and the (u,v)th element of block i is $w(t_{i,v}, t_{i,u}; \eta)$. Based on the theory of M-estimator, they derive the null distribution of Q_G by a robust inference procedure such that the LGRF test is robust to the misspecification of intrasubject correlation. As a further extension, a LGRF joint test for both marginal association and gene-time interaction was developed to integrate the time dependent genetic effect and enhance power when the gene-time interaction exists. Simulation studies [84] showed that LGRF tests can utilize the longitudinal trajectory and have improved power when compared with two existing alternatives: (1) single-marker tests using longitudinal outcome and (2) existing set-based tests using the average value of repeated measurements as the outcome.

Set-Based Tests for Gene-Environment Interaction and Future Directions in Longitudinal Studies

In search of the hidden heritability, there has been tremendous emphasis on searching for interactions between genetic factors and environmental exposures. However, examples of replicated interactions between genetic and environment factors are limited, though its role in the etiology of a disease has been broadly recognized. The findings suggest that the interaction process is probably far more complex than looking for "single locus vs. single environment factor" interaction. Following the development of set-based tests for genetic association, researchers have begun to develop methods for testing gene-environment interaction between a set of genetic variants in a gene/region and a set of correlated environmental exposures.

The first attempt is similarity regression (SIMreg) for continuous traits, which regresses phenotype similarity to genotype similarity [85]. It treats both main and interaction effects as random effects and proposes a score test. Subsequently the gene-environment set association test (GESAT) within the generalized linear model (GLM) framework was proposed by Lin et al. [86] for various types of traits (continuous, binary, etc.). It uses ridge regression to fit the main effect and proposes a variance component test for the interaction effect. It can be considered as an extension of SKAT to testing GEI. The model is

$$g(\mu_i) = \alpha_1 + \alpha_2 E_i + \sum_{k=1}^{K} G_{ik}\alpha_{3k} + \sum_{k=1}^{K} G_{ik}E_i\beta_k, \tag{8.28}$$

where g is the link function in GLM; μ_i, E_i are the mean and environmental exposure of ith subject; G_{ik} is the kth SNP of ith subject; and $\alpha_1, \alpha_2, \alpha_{3k}s, \beta_k s$ are the coefficients, respectively. We are interested in testing if there is an interaction between the set of genetic variants and the environment exposure, i.e., $H_0 : \beta = 0$. Since the number of SNPs in a region is likely to be large and some SNPs might be in high LD with each other, traditional Wald or score test can suffer from power loss and numerical difficulties. To remedy this problem, they derived a test assuming that $\beta_k s$ independently follow a distribution with mean zero and common variance τ^2. The null hypothesis $H_0 : \beta = 0$ is then equivalent to $H_0 : \tau^2 = 0$, and a variance component test statistic is proposed as

$$Q = [Y - \mu(\hat{\alpha})]^T (G * E)(G * E)^T [Y - \mu(\hat{\alpha})], \tag{8.29}$$

where $\hat{\alpha}$ is estimated under the null hypothesis: $g(\mu_i) = \alpha_1 + \alpha_2 E_i + \sum_{k=1}^{K} G_{ik}\alpha_{3k}$ using ridge regression; $G * E$ denotes the matrix of the interaction terms. GESAT inherits most advantages of a set-based score test like SKAT: It is locally most powerful under regularity conditions; it has robust power when the SNPs in the set have both positive or negative interaction with the environmental exposure; it is computationally efficient and requires only fitting the model under H_0; it can reduce the test degree of freedom and has improved power when the SNPs in the region are in high LD. In addition, they showed that single SNP analysis may be biased and has an inflated type 1 error rate under certain conditions while a region-based test can overcome this issue.

These advantages show that future developments on set based tests for GEI will be of great interest. However, opportunity coexists with new challenges. First, GEI tests may have inflated type 1 error rate when the main effect is misspecified, e.g., the true effect is quadratic but the working model is linear. Studies showed that using robust variance estimator can have robust type 1 error control, but it only works when the gene is independent of the environmental exposure [87]. The presence of this problem is more likely when approaching the interaction between marker sets, e.g., the test will be invalid as long as any one of the environmental exposure is dependent of the gene and its main effect is misspecified. Nonparametric modeling of the main effect is a plausible solution when we have a single-marker-based problem, but its feasibility is dubious when we turn to a model with hundreds of SNPs and a block of environmental exposures. Second, set-based interaction tests are more likely to suffer numerical difficulties, especially when the number of genetic variants is comparable to or larger than the sample size. It is hard even if we only fit the model under the null hypothesis, not to mention a full model with all the interaction terms.

To extend the set-based methods to a longitudinal setting, we need more efforts besides the aforementioned challenges, e.g., properly modeling the intrasubject correlation among repeated measurements; characterizing the time-varying GEI in a meaningful way. Although it requires considerable work, there is no doubt that the interaction process is of greater interest in longitudinal studies and well worth the investment, because the time-varying exposure results in more precise characterization and its interaction with genotype may further enhance the discovery process.

Analyzing Normative Aging Study Data Using Set-Based Methods

We first illustrate the earlier use of the set-based methods for genetic association to test the association between pulse pressure and 105 SNPs on genes in the iron metabolic pathway. Then we apply the set-based methods for GEI to evaluate the interaction effect between these SNPs and tibia/patella lead concentration on the pulse pressure, to assess how tibia/patella lead modify the effect of iron metabolic pathway on pulse pressure. We also include a naïve set-based method, the MinP test, which applies standard GEE to the 105 SNPs (either marginal association or interaction respectively) and adjusts the minimum p-value by calculating the effective number of tests as discussed earlier.

The kernel machine regression-based method SKAT has become popular in recent years. Since the extension of SKAT for directly analyzing repeated measures data had not been developed at the time of the analysis, we apply SKAT to the average of the longitudinal measures that can partially preserve the power attributed to analyzing longitudinal data as opposed to single time point, such as baseline, though it cannot make the best use of longitudinal information. In addition to MinP and SKAT, we also illustrate the use of LGRF, which is the most recent set-based method for a longitudinal study. The covariates fitted in the model include baseline age, time since baseline in years, and squared time. For the SKAT analysis using the average of repeated measurements, average age is used as a covariate instead. Among the three tests, MinP exhibit a p-value 0.609, SKAT is 0.279, and LGRF is 0.283. None of these showed that the marginal genetic association is significant.

Although there is no significant marginal association between the iron metabolic pathway and pulse pressure, one might be interested in whether tibia/patella lead concentration interacts with the pathway and influences pulse pressure. We apply GESAT discussed earlier 2 to the average of repeated measurements because there is no set-based GEI test for longitudinal studies, and MinP to test the interaction. The covariates adjusted are the same as the marginal association analysis. MinP test gives a significant p-value for the interaction between tibia lead and the iron pathway after adjusting for the number of effective tests (p-value = 0.042), while GESAT does not exhibit the same signal (p-value = 0.693). Neither method provides significant p-values for patella lead concentration (MinP p-value = 0.116; GESAT p-value = 0.458). Taking the average of repeated measurements of time-varying outcome (such as pulse pressure) will lead to loss of longitudinal information. This is one plausible explanation of the result difference between MinP and GESAT, because MinP based on GEE better uses the time-varying outcome in a longitudinal study but GESAT is unable to do this by reducing longitudinal data to summary averages. We also note that taking the average of repeated measurements of time-varying environmental exposure (although not showed in this analysis) is more influential for testing GEI as opposed to testing marginal association in which the genetic variants do not change over time; this latter practice may greatly reduce the power of identifying GEI in a longitudinal study. This also implies that developing a set-based GEI test for longitudinal data is of potential interest to improve the discovery process.

Characterization of Time-Varying GEI

Since phenotype traits usually vary with age, it is possible that the GEI effect has a temporal trend. For example, the detrimental effect of noise exposure on hearing threshold may depend on different genotypes among younger population but may no longer depend on genotype in older population due to age-related degeneration of hearing [88]. The attempt to characterize complex time-varying GEI effects, however, poses several statistical challenges. Prohibitive sample sizes are required when traditional models for interaction analysis are used to detect modest interactions. Furthermore, cohort studies for GEI interactions are typically characterized by substantially unequal sample size in each $G \times E \times Time$ configuration as a result of unbalanced allele-frequency, heterogeneous environmental exposure distributions in the population, and losses to follow up that are expected and could potentially bias the results in a longitudinal study. This unbalanced data structure reduces statistical power for testing any kind of time-varying pattern in the interaction parameter.

To describe potentially time-varying GEI effects, a linear mixed model is commonly used by including a $G \times E \times Time$ interaction term for possibly continuous G and E. The model may have subject-specific intercepts and slopes to represent varying changes in outcome over time across subjects. For response variable Y of the jth measurement on the ith subject Y_{ij} $(i = 1,\ldots,N, j = 1,\ldots,n_i)$, a typical linear mixed model can be postulated as,

$$Y_{ij} = \beta_0 + \beta_1 t_{ij} + \beta_2 G_i + \beta_3 E_{ij} + \beta_4 G_i E_{ij} + \beta_5 G_i t_{ij} + \beta_6 E_{ij} t_{ij} + \beta_7 G_i E_{ij} t_{ij} + b_{0i} + b_{1i} t_{ij} + e_{ij},$$

where G_i is a numeric coding of the genotype value; E_{ij} is an observed continuous environmental covariate measured at visit j; t_{ij} is the time (or age) for subject i at measurement j; b_{0i} and b_{1i} are subject specific random intercepts and random slopes, respectively; and e_{ij} is the measurement error. $E(b_{0i}) = E(b_{1i}) = E(e_{ij}) = 0$, $Var(b_{0i}) = \sigma_{b_0}^2$, $Var(b_{1i}) = \sigma_{b_1}^2$, $Var(e_{ij}) = \sigma^2$, and $Cov(b_{0i}, b_{1i}) = \sigma_{01}$. The time trend t_{ij} can be used to represent unmeasured factors that change across time [89]. Other covariates or smooth functions of variables that have nonlinear associations with Y may also be included in a GAMM framework. Model (8.28) and its extensions can be easily implemented by standard statistical software packages such as PROC MIXED in SAS or nlme in R. Here we contrast time-invariant GEI with time-varying GEI by fitting the following two models to NAS data.

$$Y_{ij} = \beta_0 + b_{0i} + \beta_{G1} G_{1i} + \beta_{G2} G_{2i} + \beta_E E_i + \beta_{G1E} G_{1i} E_i + \beta_{G2E} G_{2i} E_i + \beta_t t_{ij} + \beta_C C_i + \epsilon_{ij}$$

$$Y_{ij} = \beta_0 + b_{0i} + \beta_{G1} G_{1i} + \beta_{G2} G_{2i} + \beta_E E_i + \beta_{G1E} G_{1i} E_i + \beta_{G2E} G_{2i} E_i + \beta_t t_{ij}$$
$$+ \beta_C C_i + \beta_{Gt} G_i t_{ij} + \beta_{Et} E_i t_{ij} + \beta_{G1Et} G_{1i} E_i t_{ij} + \beta_{G2Et} G_{2i} E_i t_{ij} + \epsilon_{ij}$$

In the preceding specification, E_i is the time-invariant measure of tibia bone lead concentration for subject i. In the first model, GEI effects are represented by $\widehat{\beta_{G1E}}$ (−0.321, p-value = 0.403) and $\widehat{\beta_{G2E}}$ (0.100, p-value = 0.853). Time is a separate predictor from G and E and all the main effects and interaction effects are stationary over time. In the second model, $\widehat{\beta_{G1E}}$ (0.042, p-value = 0.942) and $\widehat{\beta_{G2E}}$ (0.684, p-value = 0.381) are the estimated GEI effects at baseline. $\widehat{\beta_{G1Et}}$ (−0.055, p-value = 0.365) and $\widehat{\beta_{G2Et}}$ (−0.133, p-value = 0.128) are the rates of increase of GEI effects over time.

Alternatively, environmental exposure can be considered as a time-varying covariate (say E_{ij}), and the interactions between gene and time-varying environment can be studied by considering lag effects [90]. The time-varying coefficient model in Liu et al. [90]

$$Y_i(t_{ij}) = \beta_0(t_{ij}) + G_{ij}\beta_g(t_{ij}) + E_{ij}\beta_e(t_{ij}) + G_{ij}E_{ij}\beta_{ge}(t_{ij}) + e_{ij}$$

allows gene effects, environment effects, and GEI interaction effects to vary over time. By adding random terms (e.g., random intercept), one can readily incorporate time-varying covariates and accommodate repeated measurements in the model while retaining the time-varying effects. A major reason why this natural extension has not been used in research applications is the lack of quality studies that collect sufficiently rich longitudinal data to make the estimation of time-varying effects reliable.

Mukherjee et al. [49] proposed an approach to capture varying time contributions to interaction terms. They utilized the AMMI model framework and constructed measures that summarize time-specific contribution to the leading interaction factors (8.10). Variation of different time contributions can be investigated by defining a contrast using the estimated factor weights $(\hat{\alpha}_r, \hat{\gamma}_c)$ [55]. Define T time intervals of a given width w to cover the study period. First calculate $Y_{trc.}$ ($t = 1, \ldots, T$), the averaged score of all observations in the rth row and cth column over the w year follow-up period. Then the variation due to time can be investigated by computing the T quantities for the mth interaction factor

$$\hat{d}_{tm} = \sum_i \sum_j \hat{\alpha}_{rm}\hat{\gamma}_{cm}Y_{trc.} \quad , t = 1, \ldots, T \tag{8.30}$$

The relative contribution of each period to interaction can be calculated by a squared term, $\left(\hat{d}_{tm} - \hat{d}_{.m}\right)^2$, where $\hat{d}_{.m}$ is $\frac{1}{T}\sum_t \hat{d}_{tm}$. The aggregate measure of squared deviation $\sum_{t=1}^{T}\left(\hat{d}_{tm} - \hat{d}_{.m}\right)^2$ captures time variability in the mth interaction term around the average regression weight across the study period. Figure 8.1 provides the graphical representation to visualize contributions of eight three-year age intervals to the first interaction factor

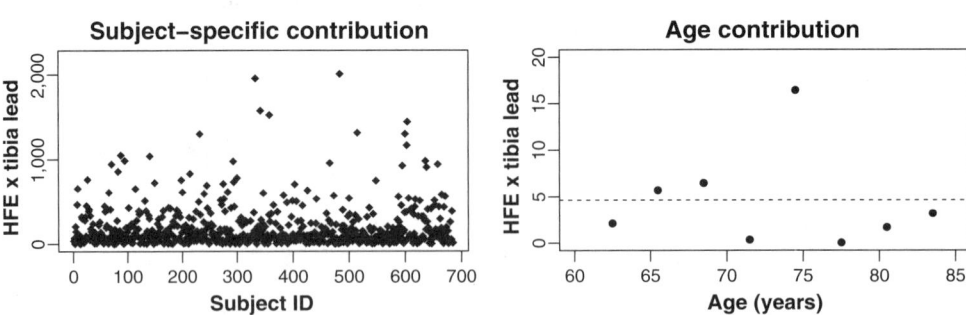

Figure 8.1
Age contributions to the first interaction factor in HFE × Lead interaction based on the Normative Aging Study data (Ko et al., 2013).

of the interaction between *HFE* gene and lead exposure through analyzing the NAS data. The plot indicates that the modifying effect of the *HFE* gene on the effect of lead exposure on PP spiked around age 75. This was due to the fact that the mean difference in PP between the Low and the High bone lead groups became largest in that age interval with H63D mutations, whereas the difference in PP among those with wild-type of HFE was the smallest.

Statistical methodology for analyzing time-varying GEI is still underdeveloped. To successfully model time-varying GEI, the models should be carefully parameterized so that they do not depend on individual's number of observations. Further development of estimation approaches with proper asymptotic theory for time-varying GEI is warranted.

Discussion

The field of studying GEI in longitudinal studies is still in its early phase of development. There are several issues that require in-depth investigation. An issue that we have completely ignored in this article is the treatment of missingness that is intrinsic to any longitudinal study. While the standard linear mixed-model analysis is valid for data missing at random, the plain vanilla GEE analysis requires data to be missing completely at random. Very few analytic choices are available to handle dropout mechanisms that cannot be ignored. For genetic association and gene-environment interaction analysis, the concerns regarding missing data remain similar to any generic longitudinal analysis. As we develop new methodology for longitudinal data, we need to devote more attention to treatment of missing data. For example, in using the score statistics in robust set-based tests, appropriate methods to handle missing data need to be developed.

Rather than test for GEI for each SNP or each gene region at a time, the combined effect of genetic variants can alternatively be converted into genetic risk scores (GRS). GRS are typically constructed by summing up the number of risk alleles on multiple markers, which may be weighted by the standardized effect size from GWAS results or other weights based on biological function and relevance. Exploring GRS × E interaction may reveal that the effects of an environmental exposure are enhanced at the extreme quantiles of GRS. GRS can be used to translate discoveries from genome-wide association studies into implications for targeted prevention for population health research [91]. How to construct a GRS based not just on marginal association of genes with disease but also GEI remains an open problem of interest.

Several papers have now shown that misspecification of main effects [87, 92] can lead to spurious detection of interactions for cross-sectional data. Similar investigations need to be carried out for longitudinal data. In addition, even if we detect a time-varying interaction, how one can biologically interpret it is a daunting task. There is some recent work on incorporating long term trajectory of exposure like body mass index in a retrospective case-control study, nested within a cohort study that are highly novel and important directions in this area [93]. To this end, design of replication studies and confirmatory functional experiments for discovered GEI is an area that requires extensive consideration. Examples of GEI studies that incorporate multiple G and multiple E are still quite limited. Latent variable models can be effectively used to handle correlated exposures in GEI studies [94]. Finally, there is something troubling about the way we conceptualize statistical interaction through various forms of product terms between G and E in a statistical model; the sensitivity of this formulation on the scale and transformation of the outcome-exposure space imply the need for more robust and scientifically plausible models for interaction that will also be able to retain adequate power. Flexible longitudinal models for the outcome-exposure relationship with potential strategies for selection of variables are a highly relevant topic for advancing our understanding of GEI relationships from a truly multivariate perspective, beyond the one at a time analysis. Similarly, design and power issues that relate variation of exposure over time, longitudinal time trend in the outcome, and range of the exposure are all important to tease apart while planning a longitudinal study. To incorporate the more complex longitudinal models into popular interfaces for doing large-scale GWAS analysis like PLINK [95] will further enhance the use of these methods by practitioners.

Acknowledgments

This research was supported by NSF grant DMS 1406712 and NIH grant ES 20811 and HL 113164.

Appendix

Table 8.4

List of the models applied to Normative Aging Study data (NAS) set

Model	Specification
G+t	$Y_{ij} = \beta_0 + b_{0i} + \beta_G G_i + \beta_t t_{ij} + \beta_{t2} t_{ij}^2 + \beta_C C_i + \epsilon_{ij}$
G*t	$Y_{ij} = \beta_0 + b_{0i} + \beta_G G_i + \beta_t t_{ij} + \beta_{Gt} G_i t_{ij} + \beta_C C_i + \epsilon_{ij}$
G*f(t)	$Y_{ij} = \beta_0 + b_{0i} + f_0(t_{ij}) + G_i f(t_{ij}) + \beta_C C_i + \epsilon_{ij}$
E	$Y_{ij} = \beta_0 + b_{0i} + \beta_E E_i + \epsilon_{ij}$
f(E)	$Y_{ij} = \beta_0 + b_{0i} + f(E_i) + \epsilon_{ij}$
G*E+t	$Y_{ij} = \beta_0 + b_{0i} + \beta_G G_i + \beta_E E_i + \beta_{GE} G_i E_i + \beta_t t_{ij} + \beta_C C_i + \epsilon_{ij}$
G*E*t	$Y_{ij} = \beta_0 + b_{0i} + \beta_G G_i + \beta_E E_i + \beta_t t_{ij} + \beta_{GE} G_i E_i + \beta_{Gt} G_i t_{ij} + \beta_{Et} E_i t_{ij} + \beta_{GEt} G_i E_i t_{ij} + \beta_C C_i + \epsilon_{ij}$

a. G: genetic factor (H63D and C282Y), E: environmental factor (tibia bone lead concentration), t: time, C: covariate (age), Y: outcome (pulse pressure)

b. Subscript i and j refer to measures at visit j for subject

c. $b_{0i} \sim N\left(0, \sigma_b^2\right)$ and $\epsilon_i = (\epsilon_{i1}, \ldots, \epsilon_{in_i})^T \sim MVN(0, \Sigma)$ where Σ is a symmetric, positive-definite, and unstructured $n_i \times n_i$ matrix

Table 8.5

Coefficient estimates and 95 percent confidence intervals for random effects model G+t ($Y_{ij} = \beta_0 + b_{0i} + \beta_G G_i + \beta_t t_{ij} + \beta_{t2} t_{ij}^2 + \beta_C C_i + \epsilon_{ij}$) based on Normative Aging Study data (NAS) set

Terms	Coefficient Estimates	95% Confidence Interval
Age	0.030	(–0.038, 0.099)
G_1(H63D)	0.349	(–0.587, 1.285)
G_2(C282Y)	0.605	(–0.666, 1.876)
Time	0.289	(0.028, 0.551)
Time^2	–0.027	(–0.041, –0.012)

Table 8.6

Coefficient estimates and 95% confidence intervals for generalized estimating equation (GEE) model G+($Y_{ij} = \beta_0 + \beta_G G_i + \beta_t t_{ij} + \beta_{t2} t_{ij}^2 + \beta_C C_i + \epsilon_{ij}$) with first-order autoregressive working correlation based on Normative Aging Study data (NAS) set

Terms	Coefficient Estimates	95% Confidence Interval
Age	0.026	(–0.041, 0.092)
G_1(H63D)	0.454	(–0.518, 1.426)
G_2(C282Y)	0.538	(–0.742, 1.817)
Time	0.306	(0.035, 0.577)
Time^2	–0.029	(–0.044,–0.013)

Table 8.7

Coefficient estimates and 95% confidence intervals for random effects model G × t ($Y_{ij} = \beta_0 + b_{0i} + \beta_G G_i + \beta_t t_{ij} + \beta_{Gt} G_i t_{ij} + \beta_C C_i + \epsilon_{ij}$) based on Normative Aging Study data (NAS) set

Terms	Coefficient Estimates	95% Confidence Interval
Age	0.027	(–0.042, 0.096)
G_1(H63D)	0.445	(–1.040, 1.930)
G_2(C282Y)	2.453	(0.467, 4.439)
Time	–0.140	(–0.232,–0.049)
G_1 × Time	–0.011	(–0.160, 0.138)
G_2 × Time	–0.255	(–0.465,–0.045)

Table 8.8

Coefficient estimates and 95% confidence intervals for generalized estimating equation (GEE) model G × t ($Y_{ij} = \beta_0 + \beta_G G_i + \beta_t t_{ij} + \beta_{Gt} G_i t_{ij} + \beta_C C_i + \epsilon_{ij}$) with first-order autoregressive working correlation based on Normative Aging Study data (NAS) set

Terms	Coefficient Estimates	95% Confidence Interval
Age	0.024	(–0.042, 0.091)
G_1(H63D)	0.318	(–1.297, 1.932)
G_2(C282Y)	2.346	(0.358, 4.334)
Time	–0.166	(–0.266,–0.066)
G_1 × Time	0.018	(–0.146, 0.182)
G_2 × Time	–0.244	(–0.476,–0.012)

Table 8.9
Estimated degrees of freedom (EDF) and p-values for the smooth terms of generalized additive mixed model G × f(t) ($Y_{ij} = \beta_0 + b_{0i} + f_0(t_{ij}) + G_i f(t_{ij}) + \beta_C C_i + \epsilon_{ij}$) based on Normative Aging Study data (NAS) set

Terms	Estimated Degrees of Freedom (EDF)	P-values
f_0 (Wild Type)	3.145	< 0.001
f_1 (H63D vs. Wild Type)	2.000	0.737
f_2 (C282Y vs. Wild Type)	2.000	0.038

Table 8.10
Coefficient estimates and 95% confidence intervals for random effects model E ($Y_{ij} = \beta_0 + b_{0i} + \beta_E E_i + \epsilon_{ij}$) based on Normative Aging Study data (NAS) set

Terms	Coefficient Estimates	95% Confidence Interval
E	0.028	(–0.341, 0.397)

Table 8.11
Coefficient estimates and 95% confidence intervals for generalized estimating equation (GEE) model E ($Y_{ij} = \beta_0 + \beta_E E_i + \epsilon_{ij}$) with first-order autoregressive working correlation based on Normative Aging Study data (NAS) set

Terms	Coefficient Estimates	95% Confidence Interval
E	–0.015	(–0.435, 0.405)

Table 8.12
Estimated degrees of freedom (EDF) and p-values for the smooth terms of generalized additive mixed model f(E) ($Y_{ij} = \beta_0 + b_{0i} + f(E_i) + \epsilon_{ij}$) based on Normative Aging Study data (NAS) set

Terms	Estimated Degrees of Freedom (EDF)	P-values
f	3.859	0.077

Table 8.13

Coefficient estimates and 95% confidence intervals for random effects model $G \times E+t$ $(Y_{ij} = \beta_0 + b_{0i} + \beta_G G_i + \beta_E E_i + \beta_{GE} G_i E_i + \beta_t t_{ij} + \beta_C C_i + \epsilon_{ij})$ based on Normative Aging Study data (NAS) set

Terms	Coefficient Estimates	95% Confidence Interval
Age	0.028	(–0.040, 0.097)
G_1(H63D)	2.643	(–1.556, 6.842)
G_2(C282Y)	–0.270	(-6.244, 5.704)
E	0.046	(–0.423, 0.515)
Time	–0.179	(–0.256,–0.103)
$G_1 \times$ Time	–0.420	(–1.166, 0.326)
$G_2 \times$ Time	0.149	(–0.882, 1.181)

Table 8.14

Coefficient estimates and 95% confidence intervals for generalized estimating equation (GEE) model $G \times E+t$ $(Y_{ij} = \beta_0 + \beta_G G_i + \beta_E E_i + \beta_{GE} G_i E_i + \beta_t t_{ij} + \beta_C C_i + \epsilon_{ij})$ with first-order autoregressive working correlation based on Normative Aging Study data (NAS) set

Terms	Coefficient Estimates	95% Confidence Interval
Age	0.026	(–0.041, 0.093)
G_1(H63D)	3.007	(–1.433, 7.447)
G_2(C282Y)	–0.469	(-6.875, 5.937)
E	0.013	(–0.481, 0.507)
Time	–0.197	(–0.285,–0.109)
$G_1 \times$ Time	–0.469	(–1.238, 0.301)
$G_2 \times$ Time	0.174	(–0.941, 1.288)

Table 8.15

Coefficient estimates and 95% confidence intervals for random effects model $G \times E \times t$ $(Y_{ij} = \beta_0 + b_{0i} + \beta_G G_i + \beta_E E_i + \beta_t t_{ij} + \beta_{GE} G_i E_i + \beta_{Gt} G_i t_{ij} + \beta_{Et} E_i t_{ij} + \beta_{GEt} G_i E_i t_{ij} + \beta_C C_i + \epsilon_{ij})$ based on Normative Aging Study data (NAS) set

Terms	Coefficient Estimates	95% Confidence Interval
Age	0.029	(–0.040, 0.098)
G_1(H63D)	1.245	(–5.194, 7.684)
G_2(C282Y)	–1.448	(–10.520, 7.624)
E	0.037	(–0.681, 0.755)
Time	–0.157	(–0.568, 0.254)
$G_1 \times$ Time	0.228	(–0.442, 0.898)
$G_2 \times$ Time	0.431	(–0.577, 1.439)
$G_1 \times E$	–0.133	(–1.252, 0.986)
$G_2 \times E$	0.682	(–0.838, 2.202)
$E \times$ Time	0.003	(–0.069, 0.076)
$G_1 \times E \times$ Time	–0.047	(–0.167, 0.074)
$G_2 \times E \times$ Time	–0.125	(–0.302, 0.052)

Table 8.16

Coefficient estimates and 95% confidence intervals for generalized estimating equation (GEE) model $G \times E \times t$ $(Y_{ij} = \beta_0 + \beta_G G_i + \beta_E E_i + \beta_t t_{ij} + \beta_{GE} G_i E_i + \beta_{Gt} G_i t_{ij} + \beta_{Et} E_i t_{ij} + \beta_{GEt} G_i E_i t_{ij} + \beta_C C_i + \epsilon_{ij})$ with first-order autoregressive working correlation based on Normative Aging Study data (NAS) set

Terms	Coefficient Estimates	95% Confidence Interval
Age	0.026	(–0.041, 0.092)
G_1(H63D)	0.453	(–4.925, 5.831)
G_2(C282Y)	–0.715	(–10.115, 8.685)
E	–0.017	(–0.655, 0.621)
Time	–0.192	(–0.748, 0.363)
$G_1 \times$ Time	0.374	(–0.396, 1.144)
$G_2 \times$ Time	0.265	(–1.142, 1.672)
$G_1 \times E$	–0.011	(–0.894, 0.873)
$G_2 \times E$	0.534	(–1.034, 2.102)
$E \times$ Time	0.005	(–0.093, 0.104)
$G_1 \times E \times$ Time	–0.069	(–0.207, 0.070)
$G_2 \times E \times$ Time	–0.093	(–0.349, 0.163)

References

1. Robertson, S. P., & Poulton, R. (2008). Longitudinal studies of gene-environment interaction in common diseases-good value for money? *Novartis Foundation Symposium*, *293*, 128–137; discussion 138–142, 181–183.

2. Bell, B., Rose, C., & Damon, A. (1966). The Veterans Administration longitudinal study of healthy aging. *Gerontologist*, *6*, 179–184.

3. Hu, H., Shih, R., Rothenberg, S., & Schwartz, B. S. (2007). The epidemiology of lead toxicity in adults: Mmeasuring dose and consideration of other methodologic issues. *Environmental Health Perspectives*, *115*(3), 455.

4. Wright, R. O., Tsaih, S. W., Schwartz, J., Wright, R. J., & Hu, H. (2003). Association between iron deficiency and blood lead level in a longitudinal analysis of children followed in an urban primary care clinic. *Journal of Pediatrics*, *142*(1), 9–14.

5. Zhang, A., Park, S., Wright, R., Weisskopf, M., Mukherjee, B., Nie, H., et al. (2010). HFE H63D polymorphism as a modifier of the effect of cumulative lead exposure on pulse pressure: the normative aging study. *Environmental Health Perspectives*, *118*, 1261–1266.

6. Hanson, E. H., Imperatore, G., & Burke, W. (2001). HFE gene and hereditary hemochromatosis: A HuGE review. *American Journal of Epidemiology*, *154*(3), 193–206.

7. Verbeke, G., & Molenberghs, G. (2009). *Linear Mixed Models for Longitudinal Data*. Springer.

8. Fitzmaurice, G. M., Laird, N. M., & Ware, J. H. (2012). *Applied Longitudinal Analysis* (Vol. 998). John Wiley & Sons.

9. Hilbe, J. M., & Hardin, J. W. (2003). *Generalized Estimating Equations*. CRC Press.

10. Diggle, P., Heagerty, P., Liang, K. Y., & Zeger, S. (2002). *Analysis of Longitudinal Data*. Oxford University Press.

11. Laird, N. M., & Ware, J. H. (1982). Random-effects models for longitudinal data. *Biometrics*, *38*, 963–974.

12. Williams, D. A. (1982). Extra-binomial variation in logistic linear models. *Applied Statistics*, *31*, 144–148.

13. Breslow, N. E. (1984). Extra-Poisson variation in log-linear models. *Applied Statistics*, *33*, 38–44.

14. Breslow, N. E., & Clayton, D. G. (1993). Approximate inference in generalized linear mixed models. *Journal of the American Statistical Association*, *88*(421), 9–25.

15. Lindstrom, M. J., & Bates, D. M. (1990). Nonlinear mixed effects models for repeated measures data. *Biometrics, 46*, 673–687.

16. Arellano-Valle, R., Bolfarine, H., & Lachos, V. (2005). Skew-normal linear mixed models. *Journal of Data Science: JDS, 3*, 415–438.

17. Zhang, P., Song, P. X.-K., Qu, A., & Greene, T. (2008). Efficient estimation for patient-specific rates of disease progression using nonnormal linear mixed models. *Biometrics, 64*, 29–38.

18. Liang, K.-Y., & Zeger, S. L. (1986). Longitudinal data analysis using generalized linear models. *Biometrika, 73*, 13–22.

19. Zeger, S. L., & Liang, K.-Y. (1986). Longitudinal data analysis for discrete and continuous outcomes. *Biometrics, 42*, 121–130.

20. Wedderburn, R. W. (1974). Quasi-likelihood functions, generalized linear models, and the Gaussian Newton method. *Biometrika, 61*, 439–447.

21. Prentice, R. L. (1988). Correlated binary regression with covariates specific to each binary observation. *Biometrics, 44*, 1033–1048.

22. Qu, A., Lindsay, B. G., & Li, B. (2000). Improving generalised estimating equations using quadratic inference functions. *Biometrika, 87*, 823–836.

23. Hastie, T., & Tibshirani, R. (1986). Generalized additive models. *Statistical Science, 1*, 297–310.

24. Lin, X., & Zhang, D. (1999). Inference in generalized additive mixed models by using smoothing splines. *Journal of the Royal Statistical Society. Series B, Statistical Methodology, 61*, 381–400.

25. Good, I., & Gaskins, R. (1971). Nonparametric roughness penalties for probability densities. *Biometrika, 58*, 255–277.

26. Anderssen, R. S., & Bloomfield, P. (1974). Numerical differentiation procedures for non-exact data. *Numerische Mathematik, 22*, 157–182.

27. Wahba, G. (1979). Convergence rates of "thin plate" smoothing splines when the data are noisy. In T. Gassner & M. Rosenblatt (Eds.), *Smoothing Techniques for Curve Estimation* (pp. 233–245). Springer.

28. Fu, W. J. (2003). Penalized estimating equations. *Biometrics, 59*, 126–132.

29. Hastie, T., & Tibshirani, R. (1993). Varying-coefficient models. *Journal of the Royal Statistical Society. Series B. Methodological, 55*, 757–796.

30. Fan, J., & Zhang, W. (1999). Statistical estimation in varying coefficient models. *Annals of Statistics, 27*, 1491–1518.

31. Ma, S., & Song, P. X. K. (2014). Varying index coefficient models. *Journal of the American Statistical Association, 110*, 341–356.

32. Xia, Y., & Li, W. (1999). On single-index coefficient regression models. *Journal of the American Statistical Association, 94*, 1275–1285.

33. Hindorff, L. A., Junkins, H. A., Hall, P. N., Mehta, J. P., & Manolio, T. A. (2011). A catalog of published genome-wide association studies. Available from: http://www.genome.gov/ gwastudies. Accessed October 2011.

34. Browning, B. L. (2008). Presto: Rapid calculation of order statistic distributions and multiple testing adjusted p-values via permutation for one and two-stage genetic association studies. *BMC Bioinformatics, 9*, 309.

35. Cheverud, J. M. (2001). A simple correction for multiple comparisons in interval mapping genome scans. *Heredity, 87*, 52–58.

36. Nyholt, D. R. (2004). A simple correction for multiple testing for single-nucleotide polymorphisms in linkage disequilibrium with each other. *American Journal of Human Genetics, 74*, 765–769.

37. Gao, X., Starmer, J., & Martin, E. R. (2008). A multiple testing correction method for genetic association studies using correlated single nucleotide polymorphisms. *Genetic Epidemiology, 32*, 361–369.

38. Furlotte, N. A., Eskin, E., & Eyheramendy, S. (2012). Genome-wide association mapping with longitudinal data. *Genetic Epidemiology, 36*, 463–471.

39. Voorman, A., Lumley, T., McKnight, B., & Rice, K. (2011). Behavior of QQ-plots and genomic control in studies of gene-environment interaction. *PLoS One, 6* (5), e19416.

40. Levy, D., DeStefano, A. L., Larson, M. G., O'Donnell, C. J., Lifton, R. P., Gavras, H., et al. (2000). Evidence for a gene influencing blood pressure on chromosome 17 genome scan linkage results for longitudinal blood pressure phenotypes in subjects from the Framingham heart study. *Hypertension, 36*, 477–483.

41. Gong, Y., & Zou, F. (2012). Varying coefficient models for mapping quantitative trait loci using recombinant inbred intercrosses. *Genetics, 190*, 475–486.

42. de Andrade, M., Guéguen, R., Visvikis, S., Sass, C., Siest, G., & Amos, C. I. (2002). Extension of variance components approach to incorporate temporal trends and longitudinal pedigree data analysis. *Genetic Epidemiology, 22*, 221–232.

43. Hopper, J. L., & Mathews, J. D. (1982). Extensions to multivariate normal models for pedigree analysis. *Annals of Human Genetics, 46*, 373–383.

44. Wang, L., Zhou, J., & Qu, A. (2012). Penalized generalized estimating equations for high dimensional longitudinal data analysis. *Biometrics, 68*, 353–360.

45. Malzahn, D., Schillert, A., Müller, M., & Bickeböller, H. (2010). The longitudinal nonparametric test as a new tool to explore gene-gene and gene-time effects in cohorts. *Genetic Epidemiology*, *34*, 469–478.

46. Badcock, P. B., Moore, E., Williamson, E., Berk, M., Williams, L. J., Bjerkeset, O., et al. (2011). Modeling gene-environment interaction in longitudinal data: Risk for neuroticism due to interaction between maternal care and the Dopamine 4 Receptor gene (DRD4). *Australian Journal of Psychology*, *63*, 18–25.

47. Li, J. J., Berk, M. S., & Lee, S. S. (2013). Differential susceptibility in longitudinal models of gene-environment interaction for adolescent depression. *Development and Psychopathology*, *25*, 991–1003.

48. Ko, Y., Chudhuri, P., Park, S., Vokonas, P., & Mukherjee, B. (2013). Novel likelihood ratio tests for screening gene-gene and gene-environment interactions with unbalanced repeated-measures data. *Genetic Epidemiology*, *37*, 581–591.

49. Mukherjee, B., Ko, Y., VanderWeele, T., Roy, A., Park, S., & Chen, J. (2012). Principal interactions analysis for repeated measures data: Application to gene-gene and gene-environment interactions. *Statistics in Medicine*, *31*, 2531–2551.

50. Tukey, J. (1949). One degree of freedom for non-additivity. *Biometrics*, *5*, 232–242.

51. Mandel, J. (1961). Non-additivity in two-way analysis of variance. *Journal of the American Statistical Association*, *56*, 878–888.

52. Tukey, J. (1962). The future of data analysis. *Annals of Mathematical Statistics*, *33*, 1–67.

53. Kooperberg, C., & LeBlanc, M. (2008). Increasing the power of identifying gene × gene interactions in genome-wide association studies. *Genetic Epidemiology*, *32*, 255–263.

54. Murcray, C., Lewinger, J., & Gauderman, W. (2009). Gene-environment interaction in genome-wide association studies. *American Journal of Epidemiology*, *169*, 219–226.

55. Gollob, H. (1968). A statistical model which combines features of factor analytic and analysis of variance techniques. *Psychometrika*, *33*, 73–115.

56. Corsten, L., & Eijnsbergen, A. (1972). Multiplicative effects in two-way analysis of variance. *Statistica Neerlandica*, *26*, 61–68.

57. Johnson, D., & Graybill, F. (1972a). An analysis of a two-way model with interaction and no replication. *Journal of the American Statistical Association*, *67*, 862–868.

58. Johnson, D., & Graybill, F. (1972b). Estimation of σ^2 in a two-way classification model with interaction. *Journal of the American Statistical Association*, *67*, 388–394.

59. Mandel, J. (1971). A new analysis of variance model for non-additive data. *Technometrics*, *13*, 1–18.

60. Marasinghe, M., & Johnson, D. (1982). A test of incomplete additivity in the multiplicative interaction model. *Journal of the American Statistical Association*, *77*, 869–877.

61. Nocedal, J., & Wright, S. (1999). *Numerical Optimization*. Springer Verlag.

62. Gabriel, K. R., & Zamir, S. (1979). Lower rank approximation of matrices by least squares with any choice of weights. *Technometrics*, *21*, 489–498.

63. Boik, R. (1989). Reduced-rank models for interaction in unequally replicated two-way classifications. *Journal of Multivariate Analysis*, *28*, 69–87.

64. Bužková, P., Lumley, T., & Rice, K. (2011). Permutation and parametric bootstrap tests for gene-gene and gene-environment interactions. *Annals of Human Genetics*, *75*, 36–45.

65. Ko, Y., Mukherjee, B., Smith, J. A., Park, S. K., Kardia, S. L. R., Allison, M. A., et al. Testing departure from additivity in Tukey's model using shrinkage: application to a longitudinal setting. (2014). *Statistics in Medicine*. doi:.10.1002/sim.6281

66. Barhdadi, A., & Dubé, M. (2010). Testing for gene-gene interaction with AMMI models. *Statistical Applications in Genetics and Molecular Biology*, *9*, 1–27.

67. Mukherjee, B., & Chatterjee, N. (2008). Exploiting gene-environment independence for analysis of case-control studies: An empirical Bayes-type shrinkage estimator to trade-off between bias and efficiency. *Biometrics*, *64*, 685–694.

68. Chapman, J., & Whittaker, J. (2008). Analysis of multiple SNPs in a candidate gene or region. *Genetic Epidemiology*, *32*, 560–566.

69. Pan, W. (2009). Asymptotic tests of association with multiple SNPs in linkage disequilibrium. *Genetic Epidemiology*, *33*, 497–507.

70. Morgenthaler, S., & Thilly, W. G. (2007). A strategy to discover genes that carry multi-allelic or mono-allelic risk for common diseases: a cohort allelic sums test (cast). *Mutation Research. Fundamental and Molecular Mechanisms of Mutagenesis*, *615*, 28–56.

71. Madsen, B. E., & Browning, S. R. (2009). A group-wise association test for rare mutations using a weighted sum statistic. *PLOS Genetics*, *5*, e1000384.

72. Han, F., & Pan, W. (2010a). A data-adaptive sum test for disease association with multiple common or rare variants. *Human Heredity*, *70*, 42–54.

73. Li, B., & Leal, S. M. (2009). Discovery of rare variants via sequencing: Implications for the design of complex trait association studies. *PLOS Genetics*, *5*, e1000481.

74. Liu, D. J., & Leal, S. M. (2010). A novel adaptive method for the analysis of next-generation sequencing data to detect complex trait associations with rare variants due to gene main effects and interactions. *PLOS Genetics, 6*, e1001156.

75. Neale, B. M., Rivas, M. A., Voight, B. F., Altshuler, D., Devlin, B., Orho-Melander, M., et al. (2011). Testing for an unusual distribution of rare variants. *PLOS Genetics, 7*, e1001322.

76. Wu, M. C., Lee, S., Cai, T., Li, Y., Boehnke, M., & Lin, X. (2011). Rare-variant association testing for sequencing data with the sequence kernel association test. *American Journal of Human Genetics, 89*, 82–93.

77. Liu, D., Ghosh, D., & Lin, X. (2008). Estimation and testing for the effect of a genetic pathway on a disease outcome using logistic kernel machine regression via logistic mixed models. *BMC Bioinformatics, 9*, 292.

78. Zhang, D., & Lin, X. (2003). Hypothesis testing in semiparametric additive mixed models. *Biostatistics, 4*, 57–74.

79. Ionita-Laza, I., Buxbaum, J. D., Laird, N. M., & Lange, C. (2011). A new testing strategy to identify rare variants with either risk or protective effect on disease. *PLOS Genetics, 7*, e1001289.

80. Basu, S., & Pan, W. (2011). Comparison of statistical tests for disease association with rare variants. *Genetic Epidemiology, 35*, 606–619.

81. He, Z., Zhang, M., Zhan, X., & Lu, Q. (2014). Modeling and testing for joint association using a genetic random field model. *Biometrics*. doi:.10.1111/biom.12160

82. Yang, J., Lee, S. H., Goddard, M. E., & Visscher, P. M. (2011). GCTA: A tool for genome-wide complex trait analysis. *American Journal of Human Genetics, 88*, 76–82.

83. Li, M., He, Z., Zhang, M., Zhan, X., Wei, C., Elston, R. C., et al. (2014). A generalized genetic random field method for the genetic association analysis of sequencing data. *Genetic Epidemiology, 38*, 242–253.

84. He, Z., Zhang, M., Lee, S., Smith, J. A., Guo, X., Palmas, W., et al. (2015). Set-based tests for genetic association in longitudinal studies. *Biometrics, 71*, 606–615.

85. Tzeng, J. Y., Zhang, D., Pongpanich, M., Smith, C., McCarthy, M. I., Sale, M. M., et al. (2011). Studying gene and gene-environment effects of uncommon and common variants on continuous traits: A marker-set approach using gene-trait similarity regression. *American Journal of Human Genetics, 89* (2), 277–288.

86. Lin, X., Lee, S., Christiani, D. C., & Lin, X. (2013). Test for interactions between a genetic marker set and environment in generalized linear models. *Biostatistics, 14*, 667–681.

87. Voorman, Arend, et al. (2011). Behavior of QQ-plots and genomic control in studies of gene-environment interaction. *PloS One, 6*(5), e19416.

88. Bainbridge, K. E., Hoffman, H. J., & Cowie, C. C. (2008). Diabetes and hearing impairment in the United States: Audiometric evidence from the National Health and Nutrition Examination Survey, 1999 to 2004. *Annals of Internal Medicine, 149*(1), 1–10.

89. Moreno-Macias, H., Romieu, I., London, S. J., & Laird, N. M. (2010). Gene-environment interaction tests for family studies with quantitative phenotypes: A review and extension to longitudinal measures. *Human Genomics, 4*, 302–326.

90. Liu, C.-Y., Maity, A., Lin, X., Wright, R. O., & Christiani, D. C. (2012). Design and analysis issues in gene and environment studies. *Environmental Health, 11*, 93.

91. Belsky, D. W., Moffitt, T. E., Sugden, K., Williams, B., Houts, R., McCarthy, J., et al. (2013). Development and evaluation of a genetic risk score for obesity. *Biodemography and Social Biology, 59*, 85–100.

92. Tchetgen, E., & Kraft, P. (2011). On the robustness of tests of genetic associations incorporating gene-environment interaction when the environmental exposure is mis-specified. *Epidemiology (Cambridge, Mass.), 22*(2), 257–261.

93. Wei, P., Tang, H., & Li, D. (2014). Functional logistic regression approach to detecting gene by longitudinal environmental exposure interaction in a case-control study. *Genetic Epidemiology, 38*(7), 638–651.

94. Sánchez, B. N., Kang, S., & Mukherjee, B. (2012). A latent variable approach to study gene-environment interactions in the presence of multiple correlated exposures. *Biometrics, 68*(2), 466–476.

95. Purcell, S., Neale, B., Todd-Brown, K., Thomas, L., Ferreira, M. A. R., Bender, D., et al. (2007). PLINK: A toolset for whole-genome association and population-based linkage analysis. *American Journal of Human Genetics, 81*, 559–575.

9 Cumulative Gene × Environment Influences on Neurobehavioral Development in Humans and Nonhuman Primates

Fatima Umber Ahmed and Erin Loraine Kinnally

The experience of stress in the form of neglect or maltreatment in childhood poses ongoing and multifaceted challenges for the developing child [1, 2]. These challenges are compounded by long-lasting abnormalities in multiple physiological systems that regulate emotional and social development. There is variation, of course, in how individuals respond to early trauma. Some of this variation may be explained by structural and functional genomic variability among individuals. Disentangling the roles of genomic and environmental influences on neurobehavioral development will help us understand and treat individuals at risk for developing lifelong psychopathology. This effort is augmented by experimental research in animal models, particularly in nonhuman primates that exhibit similar genetic variability with humans. Understanding the developmental processes and mechanisms underlying complex gene-environment interactions across species will further our understanding of the ontogeny and inheritance of normal, pathological, and resilient functioning.

Gene-Environment Interactions

In humans, childhood maltreatment can include emotional, physical, and sexual abuse, as well as emotional and physical neglect [3–5]. These types of relationships may lack the security, emotional support, and attachment cues needed for healthy development [6, 7]. A major psychopathological outcome of dysfunction in the caregiving system early in life is an increased risk for developing major depressive disorder (MDD) [8]. This risk, however, may depend on the genetic makeup of the individual. The last decade has seen a surge in scientific research demonstrating protective or risky effects of structural genetic polymorphisms in candidate genes, which has offered new understanding of risk and resilience [8–10]. Perhaps not surprisingly, the genes seemingly involved with risk and resilience are those integral to systems that regulate physiological stress responses.

One of the best-characterized gene variants that confers risk for MDD following early stress is the serotonin transporter (*5-HTT*) gene. Serotonin neurosignaling is critical in the organization and maintenance of emotion-related neural circuits, and has been strongly implicated in the neurodevelopment of psychiatric illnesses, especially MDD [8, 11]. Serotonin neurosignaling is regulated in part by the serotonin transporter (encoded by the serotonin transporter, *5-HTT,* gene), which is responsible for reuptake of serotonin from the synaptic cleft after neurotransmitter release. In humans, structural variation in the promoter of the *5-HTT* gene polymorphic region (*5-HTTLPR*) yields two different alleles in the population [11, 12]. The long and short variants may function differently, since carriers of the short allele exhibit less 5-HTT expression in blood and brain [13]. The short variant has been associated with enhancing depression susceptibility following interaction with stressful events [8, 11, 14] and the long variant may protect against stressful events [3, 8, 15–17]. The role of the serotonin transporter gene in individual risk following stressful experiences provides a specific example of the gene-environment interactions that guide neurobehavioral development and risk for psychopathology. However, one meta-analysis has recently suggested that, across different study populations, the magnitude of effect of *5-HTTLPR* may be smaller than previously thought [18]. This may not be surprising, given fact that many genes, including *5-HTT,* likely regulate response to stressful experiences. Thus, recent research has expanded the search for multigene-environment interactions in an attempt to understand the more complex dynamics of human genetics, environment, and disease.

Multigene Environment Interactions

For example, research has shown that the effects of *5-HTTLPR* may depend on the contribution of other genotypes. A common single-nucleotide polymorphism (SNP) in the gene that encodes brain-derived neurotrophic factor (*BDNF*) includes a substitution of the amino acid methionine (Met) for valine (Val) at codon 66 (Val66Met). This genotype moderates the association of *5-HTTLPR* and early abuse with depression [4, 19, 20]. Depending on the *BDNF* background (Val/Val versus Met allele), the short variant of *5-HTTLPR* conferred differential risk for depressive symptoms [4, 20]. Once again, the short variant of the *5-HTTLPR* was associated with an increased risk for depression, but only in combination with the Val/Val *BDNF* genotype. [4] These interactions were found in subjects who reported at least moderate to severe emotional, sexual or physical childhood abuse [4]. In contrast, the short *5-HTTLPR* variant was associated with protective affects against depression in carriers of Met *BDNF* that were exposed to childhood abuse [4].

The mechanisms of interaction between the *5-HTTLPR* and *BDNF* genes are not known, but we might speculate on the molecular bases of interaction. Of the four different neurotrophins that induce the survival, development, and function of neurons, BDNF has been studied most extensively. Consistent with the neurotrophic hypothesis of depression [21, 22], stress in the form of childhood adversity can lead to brain atrophy, loss of neurons and connections [23], possibly due in part to modified BDNF levels or function [24]. The *BDNF* Met allele-associated protective mechanism could operate especially during brain development and protect against the vulnerability to increased long-term risk for depressive disorders in short *5-HTTLPR* allele carriers [4, 19]. Substantial evidence from brain imaging studies and genetic studies suggest that the *BDNF* Met allele promotes the development and function of neurons, including serotonin neurons [20, 25]. Correspondingly, gray matter volume reductions are seen in brain areas implicated in depression in *5-HTTLPR* short variant carriers with a *BDNF* Val/Val genotype [20]. Further, in carriers of the short *5-HTTLPR* variant, significantly more structural connectivity is observed in carriers of the Met allele genotype than the Val/Val genotype, suggesting that the Met allele of *BDNF* may promote neurogenesis or protect against neuronal death [20, 24]. It is possible that if an individual experienced childhood maltreatment and carried the short variant of *5-HTTLPR*, the *BDNF* Met allele may confer protective effects via enhanced neuron survival or function.

Genes that regulate the hypothalamic-pituitary- adrenal (HPA) axis, an important aspect of the physiological response to stress, may also interact with serotonin-related genes to influence risk for psychopathology following stress. One important part of the HPA response is regulation of corticotropin releasing hormone (*CRH*), and its receptors (i.e., *CRHR1*). Multiple polymorphisms have been identified in each of these genes' regulatory regions. *CRHR1* haplotypes that have been identified include the *CRHR1* TAT haplotype, the *CRHR1* TCA haplotype and the *CRHR1* rs110402 haplotype [26, 27]. In one study targeting a population of about 1,200 African Americans of low socioeconomic status, with high rates of childhood and lifetime trauma, Ressler et al. (2010) discovered an interaction among *5-HTTLPR* genotype, *CRHR1* haplotype, and level of child abuse history [27]. They used the childhood trauma questionnaire, which assessed three types of childhood abuse: (sexual, physical, and emotional) to formulate a three-level abuse variable. The three-level abuse variable consisted of: (1) those with no type of abuse in the moderate-to-severe range, (2) those with moderate-to-severe on at least one type of abuse, (3) those with moderate-to-severe abuse in two or more types of abuse [27]. The *5-HTTLPR* S allele conferred risk for greater depression scores (according to the Beck Depression Inventory) in carriers of *CRHR1* risk haplotypes only in those that experienced one type of moderate to severe abuse.

Cicchetti and colleagues (2011) also demonstrated a gene-gene-environment interaction involving *CRHR1, 5-HTTLPR*, and childhood maltreatment, but the effects were on a different psychological outcome than the previously described study [26]. They examined internalizing symptomology in children, hypothesized to be a predictor of later depression. They used a combination of the Children's Depression Inventory and the Teacher Report Form as an index of high depressive and internalizing symptomology in children [26]. Interestingly, they found that maltreated children with two copies of the *CRHR1* risk haplotype and the long *5-HTTLPR* allele scored significantly higher for internalizing symptoms. These results differ from previous studies because the long *5-HTTLPR* allele was associated with risk. Nonetheless *5-HTT* and *CRH* genes appeared to interact to affect depression-related outcomes.

In analyzing the mechanisms of gene-gene interactions, Ressler and colleagues (2010) noted that early adverse experience impacted both the CRH and 5-HT systems and is likely due to the strong interconnection between the two neural circuits [27]. *CRHR1* is a G-protein-coupled receptor localized in frontal cortical areas, forebrain, brainstem, amygdala, cerebellum, and the anterior pituitary [28]. *CRHR1* plays a key role in the regulation of the HPA axis in response to stressful events, mediating in part the action of corticotropin-releasing hormone (CRH) on the pituitary to release adrenocorticotropic hormone that stimulates the production of cortisol in the adrenal cortex [29, 30]. Intriguingly, the degree of HPA response to stress is predicted by neural 5-HTT expression, one example of the correlational relationship between the two systems' functions [31].

5-HTTLPR has also been examined in conjunction with another serotonin-related genotype, a polymorphism in the promoter of the monoamine oxidase A gene (MAOA-uVNTR). Monoamine oxidase-A (MAOA) is a mitochondrial enzyme that is expressed predominantly in monoaminergic neurons and responsible for the degradation of neurotransmitters including, dopamine, serotonin, and norepinephrine [10, 32, 33]. The monoamine oxidase A-untranslated variable nucleotide tandem repeat polymorphism (MAOA-uVNTR) is located on the X chromosome and is characterized by an upstream variable number of tandem repeats (uVNTR) [10, 32–34]. In humans the MAOA-uVNTR polymorphism consists of an upstream variable number of tandem repeats (uVNTR) including 2, 3, 3.5, 4, or 5 18 base-pair repeats [32, 34]. The length of the polymorphism, which is characterized by the low vs. high number of tandem repeats, may determine in part the efficiency with which MAOA functions within individuals [10, 35, 36].

Low-activity MAOA-uVNTR alleles, (3.5 and 4 repeats in humans), have been demonstrated to confer lesser transcriptional efficiency [32] and are associated with a greater risk for aggressive and impulsive behavior in some contexts [35]. Cicchetti and colleagues (2007) investigated an interaction among childhood maltreatment, and both 5-HTTLPR and MAOA-uVNTR polymorphisms in predicting depressive symptomology in adolescents [10]. They found that the youth who had experienced child maltreatment in the form of sexual abuse and possessed both the s/s 5-HTTLPR genotype and the low activity MAOA-uVNTR genotype exhibited higher depressive and anxiety symptom scores than other groups. In contrast, sexually abused adolescents who possessed the other 5-HTT and MAOA variants were somewhat protected from their abusive experiences [10].

Gene-Gene-Environment Interactions in Nonhuman Primates

Rhesus macaques (Macaca mulatta) are one of the best translational species for investigating the effects of multiple genes and early stress on neurobehavioral development. They show genetic, neural, and social complexity that is more directly comparable to the humans than other mammalian species widely available for study [37, 38]. In rhesus macaques, one of the most commonly used experimental early life stressors is maternal deprivation (called nursery- or peer-rearing, NR), which is often compared with semi-naturalistically mother-reared subjects (NC) that live with their mothers in extended social groups, much like those observed in the wild [39, 40]. NR infants are typically removed from their mothers on the first day of life and sometimes housed in an incubator for one month to maintain infant body temperature before being pair-housed with peers of the same age. Compared with NC animals, NR macaques are characterized by widespread dysregulation of stress pathway genes [41, 42], hypothalamic-pituitary-adrenal (HPA) axis dysregulation [39, 40], and emotion dysregulation [2], similar to humans [1, 5, 43–45]. It is believed that this dysregulation arises from the absence of the developmentally important mother-infant bonding period.

Gene-gene-environment interactions have been observed in rhesus macaques, which show orthologous functional polymorphisms in the *5-HTT* and *MAOA* promoters [12, 46]. As in humans, *5-HTT* and *MAOA* gene polymorphisms have been shown to be related to psychological dysfunction in macaques who experience early adversity. For example, Kinnally and colleagues (2010) investigated the combined influence of these two serotonin pathway polymorphisms, *rh5-HTTLPR* and *rhMAOA-LPR*, on two dimensions of behavioral stress response on infant rhesus macaques in the context of nursery rearing vs. control rearing [34]. Carrying two risk genotypes was associated with greater behavioral disinhibition (lower emotional reactivity and greater activity) in NR infants, compared with control infants or NR infants with no risk genotypes.

The behavioral consequences of such gene-gene-environment interactions in macaques is consistent with work by Wendland and colleagues (2006) examining the phylogenetic relationship between social structure and *rh5-HTTLPR* and *rhMAOA-LPR* variation in seven species of the genus *Macaca*. Species characterized by more egalitarian social structures were typically monomorphic for both the *5-HTTLPR* and *MAOA-LPR* gene [47]. In contrast, macaque species characterized by rigid hierarchical structure and more aggressive behavior were polymorphic and had two or more allelic variations of *rh5-HTTLPR* and *rhMAOA-LPR* genotypes. Intriguingly, rhesus macaques, arguably the most intolerant and hierarchical species of macaques, showed the greatest degree of allelic variation in both genes [47]. These data suggest that variation in these gene regulatory regions may contribute to species-typical socioemotional traits and associated behavior.

The findings of Brent and colleagues (2013) support the notion that these and other serotonin-related genotypes may interact to influence social behavior in macaques [48]. This study illustrated that the short *5-HTTLPR*, in tandem with a variant in the regulatory region of the tryptophan hydroxylase 2 (*TPH2*, the gene that codes for the rate limiting enzyme in serotonin production) gene [49] displayed less grooming within social groups, and occupied less socially integrated positions in the social network [48]. Consistent with previous studies, it is possible that these potentially adverse social outcomes arise from emotion dysregulation in response to challenging conditions. These studies highlight that gene-gene-environment interactions are somewhat conserved in primates.

Nonlinearity in Environment-Gene-Gene Interactions

These and other genotypes have been implicated in risk for poor mental health outcomes following early life stress, but their effects have not always been demonstrated to be additive. Other candidate gene polymorphisms that have been evaluated as moderators of the effects of adversity include variants in the oxytocin receptor gene (*OXTR*), dopamine receptor genes (*DRD2* and *DRD4*), and catechol-O-methyltransferase gene (*COMT*). The *OXTR* rs53576 single nucleotide polymorphism (SNP) includes the substitution of an A allele (i.e., the minor allele) for the G allele (i.e., the major allele [50]). Previous studies examining loneliness and the *OXTR* gene in adults have shown that the AA rs53576 genotype are at greater risk for reporting loneliness than carriers of the G allele [51]. One study in adolescents explored a possible relationship between the *OXTR*, *5-HTTLPR,* and *DRD2* polymorphisms with loneliness. Van Roekel et al. (2013) did not find a direct relationship between *OXTR* and loneliness; however, an *OXTR* by sex interaction was observed [52]. Girls showed a steeper decline in loneliness when they had an A allele compared with girls who were homozygous for the G allele. A gene-gene interaction was also observed. Both boys and girls who had at least one A1 allele for the *DRD2* gene and also had the GG genotype for the *OXTR* gene showed stable levels of loneliness over time, whereas adolescents with other combinations of genotypes experienced a decrease in loneliness. The results of this study are inconsistent with previous studies that identified the A allele to be the risk variant of the *OXTR* gene [53], but consistent with others [54]. Furthermore, no interaction was found between *5-HTTLPR* and *OXTR* genotypes. Thus, it is not necessarily the case that "risk" genotypes are risky in every context.

Another example of the nonadditivity of risk genotypes is presented in Luijk et al. (2011), which tested cumulative main and interaction effects of the "risk" alleles for candidate genes involving dopamine, serotonin and oxytocin systems (*DRD4*, *DRD2*, *COMT*, *5-HTT*, *OXTR*) on attachment security and disorganization in two birth cohorts of more than 1,000 infants in total [55]. In this large cohort study, no consistent evidence emerged for additive effects of candidate genes involved in attachment security and disorganization. Breastfeeding at six months, in conjunction with genotype and maternal sensitivity, was evaluated as a predictor of secure attachment. Children breastfed at six months were more secure, and had more sensitive mothers. Paradoxically, short *5-HTTLPR* carriers, usually demonstrated to exhibit poorer outcomes, were often securely attached. For *COMT*, no associations with attachment emerged. No effects were found in either study for insecure or disorganized attachment in carries of the *DRD2* minor-T-(A1)-allele, *DRD4* 7-repeat, and A-allele of *OXTR*. *5-HTT* short allele carriers were more securely attached but the finding was not replicated in the second cohort.

One reason why we may not see additivity in the effects of so-called risk alleles is that the underlying polymorphism may confer sensitivity to early environments, rather than risk, per se [9]. For example, variants of *5-HTTLPR* and *MAOA-LPR* have been associated with sensitivity to both negative [8, 35, 56] and positive [41, 57] experiences in humans. In rhesus macaques, there is some support for the genetic plasticity hypothesis, although the low-risk alleles seemed to confer greater sensitivity. Sullivan et al. (2011) measured the strength of similarities in temperament between mothers and offspring among rhesus macaques [58]. Offspring with low risk versions of either the *rhMAOA-LPR* or the *rh5-HTTLPR* had temperament factor scores (including Vigilant, Gentle, and Nervous temperament dimensions) that were significantly predicted by their mother's temperament scores [58]. This indicates that individuals with the "low risk" forms of these alleles, especially the *5-HTTLPR*, may be more sensitive to their mother's personality, resulting in increased associations between mother and offspring temperament.

Gene Expression Networks and Neurobehavioral Development

Another approach to help us understand the role of multiple genes in behavioral development following early stress is to take a functional approach, and consider the systemwide regulation of genes in response to early life stress. Taking a dynamic gene expression approach may augment our understanding about the role of multiple genes and environment in neurobehavioral development. One of the recently identified consequences of early life stress is the reorganization of stress pathway gene expression patterns. There is substantial evidence that plasticity in genomic *activity* (production of RNA and protein) following early life stress is widespread across humans and nonhuman primates. Gene expression changes in the neural stress circuit and in blood have been reported following childhood stress in multiple species in the stress-responsive monoamine [41, 42, 44, 59], glucocorticoid [60–62], and neuropeptide [63] systems. Specifically, nonhuman primate studies have demonstrated that multiple stress pathway genes are changed following different types of early life stress. Blood and limbic brain *5-HTT* and *5-HT1A* activity is reduced in macaques that are "peer reared" [41, 42, 64]. Similarly, early repeated separation results in reduced hippocampal glucocorticoid receptor (*GR*) and *5-HT1A* in marmoset monkeys (*Callithrix jacchus*) [60, 65].

The effects of early stress on the regulation of multiple stress response genes are just beginning to be identified [59, 66, 67]. Interconnected groups of genes (or gene networks) may therefore interact to influence the neurocircuitry of the stress response. Some studies have demonstrated the coordinated expression of multiple genes in one specific pathway. For example, *GR* and *NFKB* are stress responsive transcription factors, and one study demonstrated that adults experiencing chronic caregiver stress showed differential expression of multiple gene transcripts with *GR* and *NFKB* response elements [68]. Since *GR* and *NFKB* influence the expression of other stress pathway genes in their role as transcription factors, changed expression in these two genes following stress may have wide-reaching consequences. Although this study ruled out differential *GR* expression as an explanation, it demonstrates the consequences of stress on potentially coregulated genes. Studies like these have examined coordinated gene expression patterns across all known genes using microarray technology, which quantifies the degree of expression of most known gene transcripts, or RNA molecules. The disadvantages of this approach include the expense and the need for advanced statistical analysis due to the size of the resulting dataset. Because of these disadvantages, however, the scientific community has prioritized the creation of publicly available databases for the deposition of microarray data across species. One study using publicly available microarray data identified several large families of coexpressed genes that are observed across eukaryotes [66]. It was observed that greater variance in this expression network in the hippocampus was associated with reduced stress response in mice. This study was one of the first to demonstrate a relationship between early stress, large-scale organization of gene expression patterns, and behavioral traits.

Taking a larger scale approach to understanding the role of genome-wide reorganization following early stress and its association with neurobehavioral development may be advantageous [69]. Advances in statistical methods tailored to high-throughput gene expression data have made candidate gene network analysis possible. For example, weighted Gene Co-Expression Network Analysis (WGCNA) detects groups of coexpressed genes, distinguishes groups based on connectivity strength and specific gene membership, and determines whether differences are attributable to "hub," or highly interconnected genes, strength of gene interconnections, or gene membership [70]. A recent paper demonstrated an effect of uncontrollable foot shock stress to disrupt large interconnected plasticity gene networks (including *BDNF* and *CREB* among others) in rats [71]. Another study demonstrated that hub genes within gene networks were more likely to be differentially expressed in major depressive disorder (MDD) in humans, suggesting that identifying these hub genes within gene networks may help us understand the central risk factors for MDD [72]. Notably, gene expression networks in different brain areas may be more related to psychopathology than others: another study from this group study demonstrated that the expression of a conserved set of genes related to depression in humans and in a mouse model of depression were dysregulated in the amygdala but not the anterior cingulate gyrus [73].

To explore the utility of such approaches for understanding the complexity of multigene interactions in neurobehavioral development in nonhuman primates, we used publicly available gene expression (microarray) data from the Allen Brain Institute Developing Macaque Brain Atlas to illustrate how gene expression networks are regulated in different stress-responsive brain regions across an early sensitive period in infant macaques [74]. We focused on gene expression networks in the amygdala, hippocampus, and medial prefrontal cortex because these brain areas have all been demonstrated to be instrumental in the processing of early experiences, perhaps especially early stress [62, 75–78], suggesting that these are useful target areas for understanding how early experiences reorganize lifelong behavioral development. Macrodissections of these neuroanatomical areas were collected from three newborn and three three-month-old male rhesus macaques as described in [74].

Though microarray data for thousands of genes are available for each brain region from the Allen Brain Institute resource, we focused on three narrower classes of genes (see figure 9.1 for gene list) to interrogate potential stress response gene networks. The three classes of genes were: (1) stress responsive genes, or genes that have been previously demonstrated in the literature to change in response to stress and to play a role in behavioral development across species; (2) representative transcription factors for these stress response genes; and (3) genes related to cell regulation. Our hypothesis was that gene networks would emerge among stress response genes and their transcription factors, while control cell regulation genes may be less involved with stress-response gene networks.

To select stress responsive genes in an unbiased fashion, we searched PubMed (300 hits) and Google Scholar (biological and life sciences: 20,900 hits) for all references to keywords "mammal," "gene," "stress," "environment" "brain," "development," and "behavior." We excluded review papers, considering only primary data reports. Further, because our purpose is to understand the regulation of genes involved with general psychological stress, we included only papers in which psychological stress was considered: papers that used oxidative, chemical, temperature/ salinity or endoplasmic reticulum stress were excluded. We compiled a list of all genes referenced at least twice among these publications, generating a list of thirty two candidate genes. Not surprisingly, most of these thirty two genes are known play an important role in the neural stress circuit systems, many of which we described in previous sections.

Next, we searched for stress-related transcription factors by doing analogous searches for the same keywords, adding the word "transcription factor" (15,400 hits). Intriguingly, all of the eight transcription factors identified to be stress responsive are known to regulate many of our selected candidate

GENE	FUNCTION	REFERENCES
STRESS RESPONSIVE GENES		
5-HTT	presynaptic serotonin uptake	Caspi et al., 2003
5-HT1A	serotonin receptor	Ichise et al., 2006
MAOA	serotonin, dopamine catalysis	Caspi et al., 2002
TPH2	serotonin production	Gardner et al., 2009
DRD2	dopamine receptor	Wiebe et al., 2002
DRD4	dopamine receptor	Kranenburg et al., 2009
DAT	presynatpic dopamine uptake	Bailey et al., 2010
ADRA2	postsynaptic norepinephrine receptor	Mead et al., 2010
OPRM1	opioid receptor	Kademnian et al., 2002
GR	glucocorticoidreceptor, TF	Francis et al., 1999
CRFR1	corticotropin receptor	Grabe et al., 2010
POMC	progenitor of ACTH	Cullinan et al., 1995
FKBP5	modulates GR action	Binder et al., 2008
BDNF	brain-derived neurotrophic factor	Roth et al., 2009
OXTR	oxytocin receptor	Thompson et al., 2011
AVPR1A	vasopressin receptor	Lukas et al., 2010
ER-beta	estrogen receptor, TF	Champagne et al., 2006
NPY	neuropeptide Y	Mercer et al., 1996
DNMT1	DNA methylation patterning	LaPlant et al., 2011
DNMT3A	DNA methylation patterning	LaPlant et al., 2011
DNMT3B	DNA methylation patterning	Covington et al., 2010
DNMT3L	DNA methylation patterning	LaPlant et al., 2011
HDAC 2	compacts histones surrounding DNA	Covington et al., 2009
MTHFR	methionine catalysis	Devlin et al., 2010
TRANSCRIPTION FACTORS		
SP-1	TF (5-HTT, MAOA, MR, B1R, mGLUR1)	Pena et al., 2012
AP-1	TF (MAOA, DRD2)	Ladd et al., 2007
NFKB	TF (B1R, HDAC2)	Meffert et al., 2003
EGR-1	TF (DNMT3A, CRHR1, FKBP5, MTHFR)	Revest et al., 2005
CREB	TF (BDNF)	Pruunsild et al., 2011
GATA2	TF (OPRM1, MR, ERb, MAOA)	Cole et al., 2010
NGF1A	TF (GR)	Weaver et al., 2004
YY1	TF	Sotnikov et al., 2014
HSF1	TF	Takaki et al., 2006
CONTROL GENES		
EEF	Protein synthesis	Im et al., 2009
GAPDH	metabolism	Murgatroyd et al., 2009
B-ACTIN	cytoskeleton formation	Kinnally et al., 2008
C-FOS	immediate early gene	Cullinan et al., 1995
MAP1A	Microtubule formation	Mathew et al., 2002

Figure 9.1
Selected genes and their function.

genes according to the UCSC genome browser ENCODE CHIP-Seq data and previous studies [12, 79–81]. Finally, we selected six cell regulation genes, which did not necessarily appear in these searches as stress responsive, but had been previously used as housekeeping control genes due to their role in unremitting cellular processes, such as cytoskeletal regulation and metabolism.

Because of the small sample size, we could not conduct WGCNA gene network analysis. Instead, we conducted each step of the gene network analysis with SPSS statistical software, including: factor analysis to interrogate eigengene (gene network) structure, and unsupervised cluster analysis to determine whether gene networks between neural stress pathway regions change across development.

Principal components factor analysis was conducted at each age and for each brain region. One factor score was generated to reflect the gene network expression score for each age and brain region. Across ages and brain regions, this factor loaded on the majority of stress responsive candidate, transcription factor, and cell regulation gene expression values (criteria was factor loading $\geq 0.0.4$). Though not all genes loaded on this factor in all brain regions or at both ages, a minimum of 31/38 genes loaded on this factor across ages and brain regions. Most age/region gene network factors loaded the same genes. For example, NFKB failed to load significantly on four out of six gene network factors. As predicted, we found that cell regulation genes (e.g., B-actin, GAPDH) were the least likely to load on these factors, suggesting that stress-responsive candidate genes and their transcription factors are expressed in functional networks early in development. We retained nonloading genes in each factor, however, to keep the genes consistent across factors and therefore enhancing comparability among gene network factors.

We next examined whether these gene network factors differed by brain region and developmental stage using unsupervised cluster analysis (see figure 9.2). Briefly, cluster analysis measures mathematical distance between clusters of data points, or similarity between groups of variables. Relationships between variables and clusters are graphically depicted in figure 9.2 using dendrogram mapping. Each bracket of the dendrogram reflects proximity of variables in clusters in Euclidean distance. For our purposes, bracket length corresponds to similarity in gene network structure between brain regions and developmental stages. For example, clusters with brackets at a distance of one in figure 9.2 are more similar to each other than clusters with brackets at a distance of seven.

Our hierarchical cluster analysis revealed that the gene expression network was preserved across stress neural circuit regions of the amygdala, hippocampus, and medial prefrontal cortex in newborn male macaques. The gene network factor scores for each of these brain regions clustered

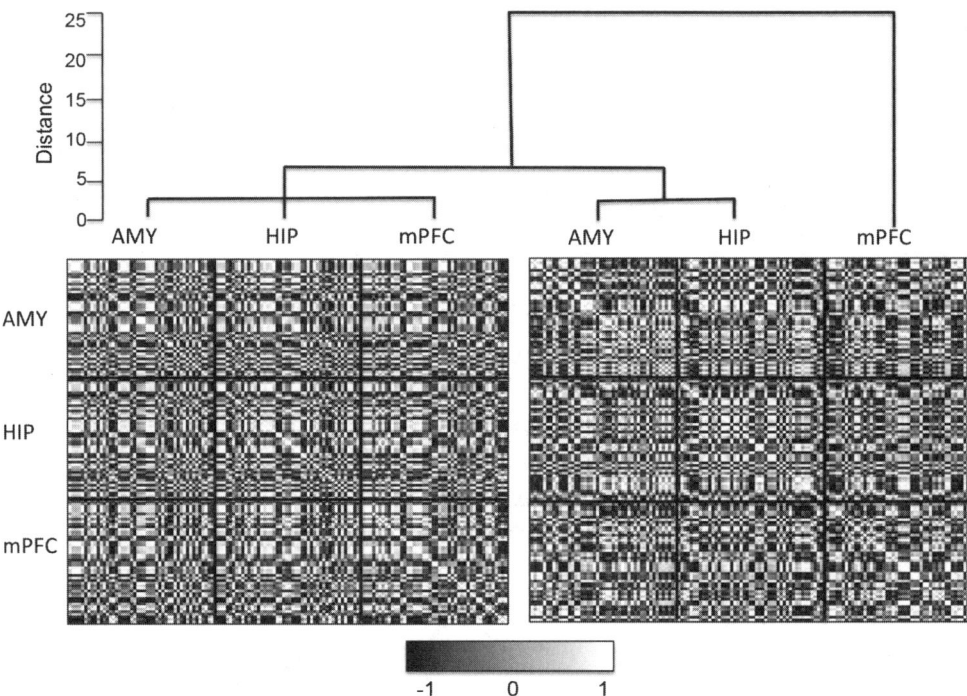

Figure 9.2

Heat map of correlations in gene expression between control, candidate, and transcription factor genes at two developmental stages in the neural stress circuit. Heat maps are depicted for 0-month-old (left) and three-month-old (right) subjects. Pearson correlation coefficients are coded between –1 (black) and 1 (white). Each cell represents the correlation coefficient between the expression of a gene represented on the X axis vs. that of another gene on the Y axis. Each gene is repeated for each tissue, delineated by blocks. Each block represents a tissue type: AMY = amygdala, HIP = hippocampus, mPFC = medial prefrontal cortex. Dendrograms denote hierarchically clustered relationships in gene networks between brain regions by age. Gene networks are preserved between brain regions at birth, but differ by three months of age, with the greatest differences seen in medial prefrontal cortex.

closely together (at a Euclidean distance of about one) at zero months of age, indicating the greatest possible degree of similarity in gene network expression scores. Similarly, gene network factor scores in the amygdala and hippocampus at three months of age clustered closely together, also indicating a degree of similarity in gene network structure and expression between these two brain regions at three months. These findings suggest that amygdala and hippocampal stress responsive gene networks may be similarly regulated across development.

Notably, however, amygdala-hippocampus gene networks differed somewhat between zero and three months. The distance between the two amygdala-hippocampus clusters (with a value of seven) was greater than the distance within clusters (at a distance of one). In short, though gene expression networks were similar between amygdala and hippocampus and zero months of age and at three months of age, the networks may change in tandem between zero and three months of age.

Gene expression in the medial prefrontal cortex shows a slightly different pattern over development. While the medial prefrontal gene expression network was similar to that in the amygdala and hippocampus at zero months of age, the greatest degree of divergence in gene networks was in the medial prefrontal cortex at three months of age. This network showed a greater magnitude of distance from both the newborn gene network cluster and the three-month-old amygdala-hippocampus gene network cluster. This suggests that the medial prefrontal cortex gene expression network is similarly regulated to amygdala and hippocampus at birth but undergoes great reorganization by three months of age.

The specifics of the relationships among gene network structure can be further interpreted by examining the degree of similarity in color pixelation between blocks in figure 9.1. This heat map presents the correlation matrices among all genes in all three brain regions at each developmental stage. The degree of correlation (ranging from −1.0 to 1.0) is reflected among all genes and brain regions using color gradation. Genes are plotted against itself and all other genes on the X and Y axis, and repeated in the same order for each block representing the amygdala, hippocampus, and medial prefrontal cortex. A correlation of −1 between two gene expression values is coded as black, a correlation of 1 is coded as white, and a correlation of zero is coded as gray. Correlation coefficients are coded as a gradient in these colors based on the magnitude of correlation. For example, a correlation of –0.5 would be represented as midway between black and gray, or dark gray. A correlation of 0.5 would be coded as midway between gray and white, or light gray. Because factor analysis relies on a correlation matrix structure, the degree of similarly in matrix color pixelation between age and brain region blocks depicts the degree of similarity in the structure of gene networks, that is, the similarly in magnitude and direction of each gene's factor loading. Examining the heat map, the difference in gene network expression factor scores across development, illustrated by hierarchical cluster analysis, appears to correspond with the structure of neural gene networks themselves, illustrated by matrix color pixelation similarity. Similar patterns of correlations among genes between brain regions in newborn monkeys are observed. Hippocampal, amygdala, and medial prefrontal cortex blocks in the newborn matrix are similar; note the clustering of white boxes in the top left corner and the interweaving black, gray, and white lines across the bottom and right edges of the blocks. In contrast, patterns of correlations between brain regions differ greatly from newborn patterns in the amygdala, hippocampus, and medial prefrontal cortex of three month olds.

This analysis is descriptive, and the samples size too small to draw statistically based conclusions. Nevertheless, the data suggest that gene networks in the stress neural circuit may include stress responsive genes, transcription factors, and, to a lesser extent, cell regulation genes. This network is relatively conserved across brain regions at birth, but is substantially reorganized between birth and three months of age in rhesus macaques. The divergence in gene networks over development may reflect specialization of gene networks by brain region at three months of age, particularly in the medial prefrontal cortex. This time period is notable because previous studies in macaques have suggested that this developmental period represents an important critical period in macaque development. Repeated maternal separation in macaques, for example, influences neurobehavioral development when administered before four months [39], but not necessarily after this age [82]. This period corresponds to the timing of infant weaning in macaque life history, and may represent a sensitive period during which gene expression networks are most susceptible to environmental disruption.

These data highlight the potential complexity in teasing apart the contribution of individual genes to stress response and neurobehavioral development. Stress responsive genes may be expressed as part of a large network of functionally interconnected genes. In studies that determine an effect of stress on single gene expression or of expression and behavior, it would be difficult to rule out a role for concomitant regulation of other genes in its network. Adding more complexity to interpretation, gene networks may function in a coordinated fashion across brain regions at birth, but may change in their strength of connectivity across early development and may develop unequally across brain regions. Thus, taking a developmental gene network approach holds great promise for advancing our understanding of the complexity of environment-gene-brain-behavior associations.

Conclusions and Future Directions

This chapter has provided examples of the complexity of gene-gene-environment interactions in guiding neurobehavioral development and risk for psychopathology in humans and nonhuman primates. We have also provided examples of functional relationships between stress-response genes and early development. Considering the effects of early stress on development not only in the context of structural genetic variation, but also in the context of gene expression variability, may lead to great advances in this field.

In the future, areas of research that examine the heritable and environmental contributions to gene expression network regulation will be essential. If we are to use gene networks to understand the effects of childhood trauma on neurobehavioral development, identifying possible mechanisms will be a critical next step. A genomic mechanism for lifelong alterations in gene expression networks and neurobehavioral development is epigenetic plasticity. The epigenome consists of at least two types of marks that play important roles in gene regulation: histone modifications and DNA methylation at CpG dinucleotides (cytosine-guanine bonded with a phosphate). DNA methylation, which typically represses gene expression, is an exciting candidate mechanism for long-term changes in gene expression because it is relatively stable, once set. Plasticity in DNA methylation during postnatal development has been linked with neurobehavioral outcomes following early stress [61, 83]. In particular, targeting "hub" genes that have widespread potential effects on other genes within a network may be a useful place to start for epigenetic interrogation. Additionally, considering the interactions among structural variation in the genome, epigenetics, and gene expression in the context of early life stress will be a useful direction to reconcile our understanding of the constraints on genomic plasticity in neurobehavioral development.

In the postgenomic era, with our rapidly advancing understanding of the complexity of genomic and environmental contributions to neurobehavioral development, our thinking about the connections among structural variation in DNA, gene expression and psychopathology must continue to expand. This expanding perspective may be the key to unlocking the role of genetics and environment in risk for mental health disorders and even normative psychosocial functioning.

Acknowledgments

Data presented in this chapter was made available by the Allen Brain Institute via the nonhuman primate developing brain atlas in accordance with the Institute's terms of use. Website: ©2014 Allen Institute for Brain Science. NIH Blueprint Non-Human Primate (NHP) Atlas [Internet]. Available from: http://www.blueprintnhpatlas.org/.

References

1. Bowlby, J. (1951). *Maternal Care and Mental Health* (Vol. 2). World Health Organization Geneva.

2. Harlow, H. F., & Zimmerman, R. R. (1959). Affectional responses in the infant monkey; orphaned baby monkeys develop a strong and persistent attachment to inanimate surrogate mothers. *Science, 130*(3373), 421–432.

3. Brown, G. W., Ban, M., Craig, T. K. J., Harris, T. O., Herbert, J., & Uher, R. (2013). Serotonin transporter length polymorphism, childhood maltreatment, and chronic depression: A specific gene–environment interaction. *Depression and Anxiety, 30*(1), 5–13.

4. Grabe, H. J., Schwahn, C., Mahler, J., Appel, K., Schulz, A., Spitzer, C., Fenske, K., Barnow, S., Freyberger, H. J., & Teumer, A. (2012). Genetic epistasis between the brain-derived neurotrophic factor Val66Met polymorphism and the 5-HTT promoter polymorphism moderates the susceptibility to depressive disorders after childhood abuse. *Progress in Neuro-Psychopharmacology and Biological Psychiatry, 36*(2), 264–270.

5. Spinetta, J. J., & Rigler, D. (1972). The child-abusing parent: A psychological review. *Psychological Bulletin, 77*(4), 296.

6. Ainsworth, M. D. (1964). Patterns of attachment behavior shown by the infant in interaction with his mother. *Merrill-Palmer Quarterly of Behavior and Development, 10*, 51–58.

7. Cicchetti, D., & Toth, S. L. (2005). Child maltreatment. *Annual Review of Clinical Psychology, 1*, 409–438.

8. Caspi, A., Sugden, K., Moffitt, T. E., Taylor, A., Craig, I. W., Harrington, H. L., McClay, J., Mill, J., Martin, J., & Braithwaite, A. (2003). Influence of life stress on depression: Moderation by a polymorphism in the 5-HTT gene. *Science, 301*(5631), 386–389.

9. Belsky, J., Jonassaint, C., Pluess, M., Stanton, M., Brummett, B., & Williams, R. (2009). Vulnerability genes or plasticity genes. *Molecular Psychiatry, 14*(8), 746–754.

10. Cicchetti, D., Rogosch, F. A., & Sturge-Apple, M. L. (2007). Interactions of child maltreatment and serotonin transporter and monoamine oxidase A polymorphisms: Depressive symptomatology among adolescents from low socioeconomic status backgrounds. *Development and Psychopathology, 19*(04), 1161–1180.

11. Calabrese, F., Guidotti, G., Middelman, A., Racagni, G., Homberg, J., & Riva, A. M. (2013). Lack of serotonin transporter alters BDNF expression in the rat brain during early postnatal development. *Molecular Neurobiology, 48*(1), 244–256.

12. Lesch, K. P., Meyer, J., Glatz, K., Flügge, G., Hinney, A., Hebebrand, J., Klauck, S. M., Poustka, A., Poustka, F., & Bengel, D. (1997). The 5-HT transporter gene-linked polymorphic region (5-HTTLPR) in evolutionary perspective: Alternative biallelic variation in rhesus monkeys. *Journal of Neural Transmission, 104*(11–12), 1259–1266.

13. Greenberg, B. D., Tolliver, T. J., Huang, S.-J., Li, Q., Bengel, D., & Murphy, D. L. (1999). Genetic variation in the serotonin transporter promoter region affects serotonin uptake in human blood platelets. *American Journal of Medical Genetics, 88*(1), 83–87.

14. Munafò, M. R., Durrant, C., Lewis, G., & Flint, J. (2009). Gene × environment interactions at the serotonin transporter locus. *Biological Psychiatry, 65*(3), 211–219.

15. Cutuli, J. J., Raby, K. L., Cicchetti, D., Englund, M. M., & Egeland , B. (2013). Contributions of maltreatment and serotonin transporter genotype to depression in childhood, adolescence, and early adulthood. *Journal of Affective Disorders, 149*(1), 30–37.

16. Karg, K., Burmeister, M., Shedden, K., & Sen, S. (2011). The serotonin transporter promoter variant (5-HTTLPR), stress, and depression meta-analysis revisited: Evidence of genetic moderation. *Archives of General Psychiatry, 68*(5), 444–454.

17. Lavigne, J. V., Herzing, L. B. K., Cook, E. H., LeBailly, S. A., Gouze, K. R., Hopkins, J., & Bryant, F. B. (2013). Gene × environment effects of serotonin transporter, dopamine receptor D4, and monoamine oxidase A genes with contextual and parenting risk factors on symptoms of oppositional defiant disorder, anxiety, and depression in a community sample of 4-year-old children. *Development and Psychopathology, 25*(02), 555–575.

18. Risch, N., Herrell, R., Lehner, T., Liang, K-Y, Eaves, L., Hoh, J., Griem, A., Kovacs, M., Ott, J., & Merikangas, K. R. (2009). Interaction between the serotonin transporter gene (5-HTTLPR), stressful life events, and risk of depression: A meta-analysis. *JAMA, 301*(23), 2462–2471.

19. Goodyer, I. M., Croudace, T., Dudbridge, F., Ban, M., & Herbert, J. (2010). Polymorphisms in BDNF (Val66Met) and 5-HTTLPR, morning cortisol and subsequent depression in at-risk adolescents. *British Journal of Psychiatry, 197*(5), 365–371.

20. Pezawas, L., Meyer-Lindenberg, A., Goldman, A. L., Verchinski, B. A., Chen, G., Kolachana, B. S., Egan, M. F., Mattay, V. S., Hariri, A. R., & Weinberger, D. R. (2008). Evidence of biologic epistasis between BDNF and SLC6A4 and implications for depression. *Molecular Psychiatry, 13*(7), 709–716.

21. Duman, R. S., Heninger, G. R., & Nestler, E. J. (1997). A molecular and cellular theory of depression. *Archives of General Psychiatry, 54*(7), 597–606.

22. Duman, R. S., & Monteggia, L. M. (2006). A neurotrophic model for stress-related mood disorders. *Biological Psychiatry, 59*(12), 1116–1127.

23. Dranovsky, A., & Hen, R. (2006). Hippocampal neurogenesis: Regulation by stress and antidepressants. *Biological Psychiatry, 59*(12), 1136–1143.

24. Gerritsen, L., Tendolkar, I., Franke, B., Vasquez, A. A., Kooijman, S., Buitelaar, J., Fernández, G., & Rijpkema, M. (2011). BDNF Val66Met genotype modulates the effect of childhood adversity on subgenual anterior cingulate cortex volume in healthy subjects. *Molecular Psychiatry, 17*(6), 597–603.

25. Martinowich, K., & Lu, B. (2007). Interaction between BDNF and serotonin: Role in mood disorders. *Neuropsychopharmacology, 33*(1), 73–83.

26. Cicchetti, D., Rogosch, F. A., & Oshri, A. (2011). Interactive effects of corticotropin releasing hormone receptor 1, serotonin transporter linked polymorphic region, and child maltreatment on diurnal cortisol regulation and internalizing symptomatology. *Development and Psychopathology, 23*(04), 1125–1138.

27. Ressler, K. J., Bradley, B., Mercer, K. B., Deveau, T. C., Smith, A. K., Gillespie, C. F., et al. (2010). Polymorphisms in CRHR1 and the serotonin transporter loci: Gene × gene × environment interactions on depressive symptoms. *American Journal of Medical Genetics. Part B, Neuropsychiatric Genetics, 153*(3), 812–824.

28. Steckler, T., & Holsboer, F. (1999). Corticotropin-releasing hormone receptor subtypes and emotion. *Biological Psychiatry*, *46*(11), 1480–1508.

29. Kranzler, H. R., Feinn, R., Nelson, E. C., Covault, J., Anton, R. F., Farrer, L., & Gelernter, J. (2011). A CRHR1 haplotype moderates the effect of adverse childhood experiences on lifetime risk of major depressive episode in African-American women. *American Journal of Medical Genetics Part B: Neuropsychiatric Genetics*, *156*(8), 960–968.

30. Polanczyk, G., Caspi, A., Williams, B., Price, T. S., Danese, A., Sugden, K., Uher, R., Poulton, R., & Moffitt, T. E. (2009). Protective effect of CRHR1 gene variants on the development of adult depression following childhood maltreatment: Replication and extension. *Archives of General Psychiatry*, *66*(9), 978–985.

31. Reimold, M., Knobel, A., Rapp, M. A., Batra, A., Wiedemann, K., Ströhle, A., et al. (2011). Central serotonin transporter levels are associated with stress hormone response and anxiety. *Psychopharmacology*, *213*(2–3), 563–572.

32. Sabol, S. Z., Hu, S., & Hamer, D. (1998). A functional polymorphism in the monoamine oxidase A gene promoter. *Human Genetics*, *103*(3), 273–279.

33. Youdim, M. B. H., Edmondson, D., & Tipton, K. F. (2006). The therapeutic potential of monoamine oxidase inhibitors. *Nature Reviews Neuroscience*, *7*(4), 295–309.

34. Kinnally, E. L., Karere, G. M., Lyons, L. A., Mendoza, S. P., Mason, W. A., & Capitanio, J. P. (2010). Serotonin pathway gene–gene and gene–environment interactions influence behavioral stress response in infant rhesus macaques. *Development and Psychopathology*, *22*(01), 35–44.

35. Caspi, A., McClay, J., Moffitt, T. E., Mill, J., Martin, J., Craig, I. W., Taylor, A., & Poulton, R. (2002). Role of genotype in the cycle of violence in maltreated children. *Science*, *297*(5582), 851–854.

36. Kim-Cohen, J., Caspi, A., Taylor, A., Williams, B., Newcombe, R., Craig, I. W., & Moffitt, T. E. (2006). MAOA, maltreatment, and gene–environment interaction predicting children's mental health: New evidence and a meta-analysis. *Molecular Psychiatry*, *11*(10), 903–913.

37. Capitanio, J. P., & Emborg, M. E. (2008). Contributions of non-human primates to neuroscience research. *The Lancet*, *371*(9618), 1126–1135.

38. Phillips, K. A., Bales, K. L., Capitanio, J. P., Conley, A., Czoty, P. W., Hart, B. A., Hopkins, W. D., Hu, S., Miller, L. A., & Nader, M. A. (2014). Why primate models matter. *American Journal of Primatology*, *76*, 801–827.

39. Capitanio, J. P., Mendoza, S. P., Mason, W. A., & Maninger, N. (2005). Rearing environment and hypothalamic-pituitary-adrenal regulation in young rhesus monkeys (*Macaca mulatta*). *Developmental Psychobiology*, *46*(4), 318–330.

40. Higley, J. D., Mehlman, P. T., Poland, R. E., Taub, D. M., Vickers, J., Suomi, S. J., & Linnoila, M. (1996). CSF testosterone and 5-HIAA correlate with different types of aggressive behaviors. *Biological Psychiatry*, *40*(11), 1067–1082.

41. Kinnally, E. L., Huang, Y., Haverly, R., Burke, A. K., Galfalvy, H., Brent, D. P., Oquendo, M. A., & Mann, J. J. (2009). Parental care moderates the influence of MAOA-uVNTR genotype and childhood stressors on trait impulsivity and aggression in adult women. *Psychiatric Genetics*, *19*(3), 126.

42. Spinelli, S., Chefer, S., Carson, R. E., Jagoda, E., Lang, L., Heilig, M., Barr, C. S., Suomi, S. J., Higley, J. D., & Stein, E. A. (2010). Effects of early-life stress on seroto-nin$_{1A}$ receptors in juvenile rhesus monkeys measured by positron emission tomography. *Biological Psychiatry*, *67*(12), 1146–1153.

43. Gunnar, M. R., Brodersen, L., Nachmias, M., Buss, K., & Rigatuso, J. (1996). Stress reactivity and attachment security. *Developmental Psychobiology*, *29*(3), 191–204.

44. Miller, J. M., Kinnally, E. L., Ogden, R. T., Oquendo, M. A., Mann, J. J., & Parsey, R. V. (2009). Reported childhood abuse is associated with low serotonin transporter binding in vivo in major depressive disorder. *Synapse*, *63*(7), 565–573.

45. Nemeroff, C. B. (2003). Early-life adversity, CRF dysregulation, and vulnerability to mood and anxiety disorders. *Psychopharmacology Bulletin*, *38*(1), 14–20.

46. Newman, T. K., Syagailo, Y. V., Barr, C. S., Wendland, J. R., Champoux, M., Graessle, M., Suomi, S. J., Higley, J. D., & Lesch, K.-P. (2005). Monoamine oxidase A gene promoter variation and rearing experience influences aggressive behavior in rhesus monkeys. *Biological Psychiatry*, *57*(2), 167–172.

47. Wendland, J. R., Lesch, K.-P., Newman, T. K., Timme, A., Gachot-Neveu, H., Thierry, B., & Suomi, S. J. (2006). Differential functional variability of serotonin transporter and monoamine oxidase a genes in macaque species displaying contrasting levels of aggression-related behavior. *Behavior Genetics*, *36*(2), 163–172.

48. Brent, L. J. N., Heilbronner, S. R., Horvath, J. E., Gonzalez-Martinez, J., Ruiz-Lambides, A., Robinson, A. G., Skene, J. H. P., & Platt, M. L. (2013). Genetic origins of social networks in rhesus macaques. *Scientific Reports*, *3*.

49. Chen, G-L, Novak, M. A., Meyer, J. S., Kelly, B. J., Vallender, E. J., & Miller, G. M. (2010). TPH2 5′-and 3′-regulatory polymorphisms are differentially associated with HPA axis function and self-injurious behavior in rhesus monkeys. *Genes, Brain and Behavior*, *9*(3), 335–347.

50. Wu, S., Jia, M., Ruan, Y., Liu, J., Guo, Y., Shuang, M., et al. (2005). Positive association of the oxytocin receptor gene (OXTR) with autism in the Chinese Han population. *Biological Psychiatry*, *58*(1), 74–77.

51. Lucht, M. J., Barnow, S., Sonnenfeld, C., Rosenberger, A., Grabe, H. J., Schroeder, W., Völzke, H., Freyberger, H. J., Herrmann, F. H., & Kroemer, H. (2009). Associations between the oxytocin receptor gene (OXTR) and affect, loneliness and intelligence in normal subjects. *Progress in Neuro-Psychopharmacology and Biological Psychiatry, 33*(5), 860–866.

52. van Roekel, E., Verhagen, M., Engels, R. C. M. E., Goossens, L., & Scholte, R. H. J. (2013). Oxytocin receptor gene (OXTR) in relation to loneliness in adolescence: Interactions with sex, parental support, and DRD2 and 5-HTTLPR genotypes. *Psychiatric Genetics, 23*(5), 204–213.

53. Bakermans-Kranenburg, M. J., & van IJzendoorn, M. H. (2008). Oxytocin receptor (OXTR) and serotonin transporter (5-HTT) genes associated with observed parenting. *Social Cognitive and Affective Neuroscience, 3*(2), 128–134.

54. Costa, B., Pini, S., Gabelloni, P., Abelli, M., Lari, L., Cardini, A., Muti, M., Gesi, C., Landi, S., & Galderisi, S. (2009). Oxytocin receptor polymorphisms and adult attachment style in patients with depression. *Psychoneuroendocrinology, 34*(10), 1506–1514.

55. Luijk, M. P. C. M., Roisman, G. I., Haltigan, J. D., Tiemeier, H., Booth-LaForce, C., van IJzendoorn, M. H., Belsky, J., Uitterlinden, A. G., Jaddoe, V. W. V., & Hofman, A. (2011). Dopaminergic, serotonergic, and oxytonergic candidate genes associated with infant attachment security and disorganization? In search of main and interaction effects. *Journal of Child Psychology and Psychiatry, 52*(12), 1295–1307.

56. Bennett, A. J., Lesch, K. P., Heils, A., Long, J. C., Lorenz, J. G., Shoaf, S. E., Champoux, M., Suomi, S. J., Linnoila, M. V., & Higley, J. D. (2002). Early experience and serotonin transporter gene variation interact to influence primate CNS function. *Molecular Psychiatry 7*(1), 118–122.

57. Kaufman, J., Yang, B.-Z., Douglas-Palumberi, H., Grasso, D., Lipschitz, D., Houshyar, S., Krystal, J. H., & Gelernter, J. (2006). Brain-derived neurotrophic factor–5-HTTLPR gene interactions and environmental modifiers of depression in children. *Biological Psychiatry, 59*(8), 673–680.

58. Sullivan, E. C., Mendoza, S. P., & Capitanio, J. P. (2011). Similarity in temperament between mother and offspring rhesus monkeys: Sex differences and the role of monoamine oxidase-a and serotonin transporter promoter polymorphism genotypes. *Developmental Psychobiology, 53*(6), 549–563.

59. Cole, S. W., Conti, G., Arevalo, J. M., Ruggiero, A. M., Heckman, J. J., & Suomi, S. J. (2012). Transcriptional modulation of the developing immune system by early life social adversity. *Proceedings of the National Academy of Sciences of the United States of America, 109*(50), 20578–20583.

60. Law, A. J., Pei, Q., Walker, M., Gordon-Andrews, H., Weickert, C. S., Feldon, J., Pryce, C. R., & Harrison, P. J. (2008). Early parental deprivation in the marmoset monkey produces long-term changes in hippocampal expression of genes involved in synaptic plasticity and implicated in mood disorder. *Neuropsychopharmacology, 34*(6), 1381–1394.

61. McGowan, P. O., Sasaki, A., D'Alessio, A. C., Dymov, S., Labonté, B., Szyf, M., Turecki, G., & Meaney, M. J. (2009). Epigenetic regulation of the glucocorticoid receptor in human brain associates with childhood abuse. *Nature Neuroscience, 12*(3), 342–348.

62. Meaney, M. J. (2001). Maternal care, gene expression, and the transmission of individual differences in stress reactivity across generations. *Annual Review of Neuroscience, 24*(1), 1161–1192.

63. Zhao, D.-Q., & Ai, H.-B. (2011). Oxytocin and vasopressin involved in restraint water-immersion stress mediated by oxytocin receptor and vasopressin 1b receptor in rat brain. *PloS One, 6*(8), e23362.

64. Ichise, M., Vines, D. C., Gura, T., Anderson, G. M., Suomi, S. J., Higley, J. D., & Innis, R. B. (2006). Effects of early life stress on [11C] DASB positron emission tomography imaging of serotonin transporters in adolescent peer-and mother-reared rhesus monkeys. *Journal of Neuroscience, 26*(17), 4638–4643.

65. Arabadzisz, D., Diaz-Heijtz, R., Knuesel, I., Weber, E., Pilloud, S., Dettling, A. C., Feldon, J., Law, A. J., Harrison, P. J., & Pryce, C. R. (2010). Primate early life stress leads to long-term mild hippocampal decreases in corticosteroid receptor expression. *Biological Psychiatry, 67*(11), 1106–1109.

66. Alter, M. D., Rubin, D. B., Ramsey, K., Halpern, R., Stephan, D. A., Abbott, L. F., & Hen, R. (2008). Variation in the large-scale organization of gene expression levels in the hippocampus relates to stable epigenetic variability in behavior. *PLoS One, 3*(10), e3344.

67. Weaver, I. C. G., Meaney, M. J., & Szyf, M. (2006). Maternal care effects on the hippocampal transcriptome and anxiety-mediated behaviors in the offspring that are reversible in adulthood. *Proceedings of the National Academy of Sciences of the United States of America, 103*(9), 3480–3485.

68. Miller, G. E., Chen, E., Sze, J., Marin, T., Jesusa, M. G., Arevalo, R. D., Ma, R., & Cole, S. W. (2008). A functional genomic fingerprint of chronic stress in humans: Blunted glucocorticoid and increased NF-κB signaling. *Biological Psychiatry, 64*(4), 266–272.

69. Gaiteri, C., Ding, Y., French, B., Tseng, G. C., & Sibille, E. (2014). Beyond modules and hubs: The potential of gene coexpression networks for investigating molecular mechanisms of complex brain disorders. *Genes Brain & Behavior, 13*(1), 13–24.

70. Langfelder, P., & Horvath, S. (2008). WGCNA: An R package for weighted correlation network analysis. *BMC Bioinformatics*, *9*(1), 559.

71. Ponomarev, I., Rau, V., Eger, E. I., Harris, R. A., & Fanselow, M. S. (2010). Amygdala transcriptome and cellular mechanisms underlying stress-enhanced fear learning in a rat model of posttraumatic stress disorder. *Neuropsychopharmacology*, *35*(6), 1402–1411.

72. Gaiteri, C., & Sibille, E. (2011). Differentially expressed genes in major depression reside on the periphery of resilient gene coexpression networks. *Frontiers in Neuroscience*, *5*, 95.

73. Sibille, E., Wang, Y., Joeyen-Waldorf, J., Gaiteri, C., Surget, A., Oh, S., et al. (2009). A molecular signature of depression in the amygdala. *American Journal of Psychiatry*, *166*(9), 1011–1024.

74. Bernard, A., Lubbers, L. S., Tanis, K. Q., Luo, R., Podtelezhnikov, A. A., Finney, E. M., McWhorter, M. E. et al. (2012). Transcriptional architecture of the primate neocortex. *Neuron*, *73*(6), 1083–1099.

75. Sapolsky, R. M., Krey, L. C., & McEwen, B. S. (1984). Glucocorticoid-sensitive hippocampal neurons are involved in terminating the adrenocortical stress response. *Proceedings of the National Academy of Sciences of the United States of America*, *81*(19), 6174–6177.

76. Spinelli, S., Chefer, S., Suomi, S. J., Higley, J. D., Barr, C. S., & Stein, E. (2009). Early-life stress induces long-term morphologic changes in primate brain. *Archives of General Psychiatry*, *66*(6), 658–665.

77. Sullivan, R. M., & Gratton, A. (2002). Prefrontal cortical regulation of hypothalamic-pituitary-adrenal function in the rat and implications for psychopathology: Side matters. *Psychoneuroendocrinology*, *27*(1), 99–114.

78. Weidenfeld, J., Newman, M. E., Itzik, A., Gur, E., & Feldman, S. (2002). The amygdala regulates the pituitary-adrenocortical response and release of hypothalamic serotonin following electrical stimulation of the dorsal raphe nucleus in the rat. *Neuroendocrinology*, *76*(2), 63–69.

79. Wei, P., & Vedeckis, W. V. (1997). Regulation of the glucocorticoid receptor gene by the AP-1 transcription factor. *Endocrine*, *7*(3), 303–310.

80. Yao, M., & Denver, R. J. (2007). Regulation of vertebrate corticotropin-releasing factor genes. *General and Comparative Endocrinology*, *153*(1), 200–216.

81. Yoshida, M., Iwasaki, Y., Asai, M., Takayasu, S., Taguchi, T., Itoi, K., et al. (2006). Identification of a functional AP1 element in the rat vasopressin gene promoter. *Endocrinology*, *147*(6), 2850–2863.

82. Sanchez, M., Ladd, C. O., & Plotsky, P. M. (2001) Early adverse experience as a developmental risk factor for later psychopathology: Evidence from rodent and primate models. *Development and Psychopathology, 13*(03), 419–449.

83. Provençal, N., Suderman, M. J., Guillemin, C., Massart, R., Ruggiero, A., Wang, D., et al. (2012). The signature of maternal rearing in the methylome in rhesus macaque prefrontal cortex and T cells. *Journal of Neuroscience, 32*(44), 15626–15642.

10 Complex Phenotypes and the Use of Polygenic Scores to Investigate Gene × Environment Interactions for Substance Use

Michael Windle

Genetic phenotypes are typically described as either monogenic (i.e., single allele cause) or complex (i.e., multiple genetic and environmental causes). Monogenic phenotypes follow Mendel's laws of inheritance (autosomal or X-linked, dominant or recessive) and genetic disorders (e.g., cystic fibrosis, fragile X syndrome) are caused by mutations in a single gene. Complex phenotypes do not necessarily follow Mendel's laws as genes may connect with other genes and/or with environmental exposures to predict a given phenotype (e.g., diabetes, substance dependence) or intermediate phenotype (e.g., impulsivity, stress reactivity). Genes associated with complex phenotypes are often referred to as susceptibility genes and their joint contributions with other genes and environments greatly increase the conceptual, methodological, and computational challenges for investigators. This chapter focuses on complex phenotypes and the presentation and development of some conceptual and analytic issues and methods that may be used to address G × E relationships.

Initially, the most frequently used statistical approaches both for candidate gene studies and genome-wide association (GWA) studies were directed toward the investigation of single locus effects (e.g., single VNTRs or single SNPs). In GWA studies, where thousands of individual SNPs are tested, a large number of individual statistical tests are conducted that necessitate corrections to the nominal alpha level to control the type 1 error rate. For example, a common stringent alpha level such as $p < 5 \times 10^{-8}$ may be used as the threshold for genome-wide statistical significance. Such stringent thresholds for common variants that we now know typically manifest relatively small effect sizes for complex phenotypes may result in rejecting genetic loci that may be relevant to the disease or trait of interest, especially for samples of low or moderate size. Statistical power issues are compounded in G × E studies because the sample size for adequately powered tests for GE interactions need to be approximately three to four times as large as for tests of genetic main effects of comparable effect size [1, 2].

A portion of the recent biostatistics literature on GWA studies has focused on a range of methods to reduce the number of statistical tests required. One of these approaches has been to use different methods of correcting for multiple tests. The Bonferroni correction method is often used in the multiple test comparison context in which p-levels are adjusted based on the number of statistical comparisons. This correction assumes that each of the tests is independent but may be overly conservative when testing genetic loci such as SNPs that may exhibit linkage disequilibrium, i.e., a correlated dependence structure. An alternative correction method for type 1 error rates is the false discovery rate (FDR), defined as the percentage of statistical tests deemed significant that are false positives [3]. By controlling the FDR, on average, only the set alpha level (usually 0.05) of the total number of positive findings is false. Permutation testing is another method that includes multiple adjustments on the correlational structure between tests (thereby incorporating the dependence structure among units, e.g., SNPs) and significance is determined by counting the number of ways the data could be permuted to produce results more extreme than those observed [4, 5].

A second approach has been to adopt two-stage strategies that attempt to use a first-stage screening of SNPs and then a second stage analysis of a subset of SNPs meeting specified criteria (e.g., meet specified p levels); the procedure yields a smaller set of required adjustments to the nominal alpha level than standard genome-wide SNP analyses [6]. Discovery (training) and confirmation (cross-validation) samples are often used with two-stage approaches. A third approach has been to use the gene level as the unit of analysis rather than single loci such as individual SNPs [7, 8]. Statistically, using the gene level as the target instead of individual SNPs may reduce the number of statistical tests by more than tenfold, thereby reducing substantially the magnitude of alpha level corrections required to control the false discovery rate. Another gene-level approach has extended the framework to address research questions about gene level associations for multiple diseases or complex traits. Patterns of comorbidity or co-occurrence are common among many diseases and complex traits. Multiple trait tests at the gene level evaluate associations of the identified gene level score individually and jointly with the multiple traits. Guo et al. [7] demonstrated the use of this method with data from the Study of Addictions: Genetics and Environment (SAGE) dataset [9]. SAGE had identified the PKNOX2 gene as significantly associated with substance dependence. Guo et al. [7] included 13 SNPs in the PKNOX2 gene and used these with six substance-dependence disorders (alcohol, cocaine, marijuana, nicotine, opiates, and other substances) to compute associations individually for each substance as well as jointly across substances. The findings indicated greater statistical power for the multiple-trait gene-based tests relative to single-trait-based tests if common genetic variants existed across the multiple traits.

Yet another alternative approach to single locus effects has emerged in the form of multilocus and polygenic scores, which have included extensions to G × E interactions [10–12]. This approach uses a method of summing a set of either weighted or unweighted individual loci (e.g., SNPs) to form a composite index that is then used in predicting phenotypes. The number of statistical tests to be conducted is thus reduced from thousands or hundreds of individual loci to one index, thereby reducing the denominator for corrections to alpha levels. A primary focus of this chapter is on conceptual and methodological issues related to the use of polygenic scores in the study of G × E relationships.

Furthermore, for several reasons, this chapter is directed more toward addressing issues related to testing substantive G × E hypotheses rather than improving statistical methods or procedures per se. First, many complex behavioral phenotypes, including substance use, are embedded in multiple levels of the GE interplay, and GWA studies have typically not included, for example, the collection of psychosocial and environmental determinants (exposures) of health outcomes. Without suitable data on the environmental side of the equation, it is difficult to evaluate GE relationships on a scale that would benefit from atheoretical GWAS-based data mining procedures. Note that I am referring here to GE data that include both genomic material and measures of the environment and not biometrical applications that derive statistical estimates of GE relations (i.e., not standard heritability estimates based on twin designs).

Second, many behavioral-based studies contain sample sizes that are too small (and underpowered) to adequately assess thousands of GE interactions with stringent alpha corrections for multiple testing. For example, neuroimaging studies that identify key brain regions and examine associated genetic loci singly or jointly in interaction with environmental exposures are unlikely to have data on more than 100,000 participants due to feasibility issues (e.g., financial constraints). Furthermore, neuroimaging studies typically already confront challenging issues related to multiple testing corrections due to the large number of units (e.g., voxels, ROIs) involved in testing hypotheses about group differences, within-subject changes, or network connections [13]. A similar situation occurs for randomized clinical trials, which are likely to be underpowered to test an extensive set of GE relationships. A large RCT may consist of 1,000–3,000 participants in each arm of a trial, and this may be sufficiently powered statistically to test the null hypothesis regarding a treatment effect, but these numbers fall short of the sample sizes required for GWA studies. Likewise, many critical endophenotypes (e.g., electrocortical responses, physiological and hormone markers) and intermediate phenotypes (e.g., personality traits, neuropsychological dimensions of disinhibition) for which high heritability estimates exist are also unlikely to be evaluated with large scale GWAS or Next Generation Sequencing (NGS) designs, but nevertheless are clearly of high priority to current science and knowledge generation [14]. Third, for longitudinal phenotypes related to substance use progression (e.g., onset, escalation, problem use, disorder, relapse), it may be difficult to replicate findings across prospective studies because of variation across studies with regard to samples, measures, age at initial wave of measurement, intervals between measurement occasions, and so on, though new approaches (e.g., data harmonization) may provide a useful methodology to address some of these limitations [15, 16].

It is for these three reasons (and most likely others) that a theory or substantively based hypothesis strategy is likely to remain a viable (perhaps necessary) option for studies of many complex behavioral phenotypes to complement GWAS and NGS studies. Both statistically and substantively driven approaches have similar end goals of identifying critical genomic signals and associated regulatory processes (e.g., brain circuits, metabolic processes) that contribute to maladaptive behaviors and disorders or to resilience. To illustrate some of these points related to theory or to substantively based hypothesis testing, and to identify the potential value of polygenic scores to investigate GE relationships, the focal complex phenotypes used in this chapter are from the domain of substance use. The majority of the issues addressed apply similarly to other complex phenotypes (e.g., schizophrenia, obesity, diabetes).

Substance-Use Phenotypes

Globally, substance use and abuse are among the major causes of morbidity and mortality. For example, Michaud et al. [17] reported that substance use and abuse was the fifth-leading cause of disability-adjusted life years in the United States. National data in the United States also indicated a high prevalence (8.2 percent) of last year adult alcohol disorders and other substance use disorders in 2013 [18], as well as a high prevalence of early onset use and heavy episodic drinking among underage populations (e.g., children and adolescents) [19]. A plethora of behavior genetic studies have supported high heritability (e.g., 40–60 percent) for substance use disorders [20], and molecular genetic studies have identified a range of typically small effect candidate genes that are associated with SUDs, levels or intensity of substance use, and intermediate phenotypes of SUDs [14, 21]. As is the case with many GE interaction studies, studies examining candidate gene-environment relations for substance use phenotypes and intermediate phenotypes have yielded mixed results, often with difficulties replicating findings and often using underpowered research designs to adequately assess GE relationships [22–24].

Polygenic Scores

Among the strategies proposed for the advancement of studies of genomics and GE interactions has been the adoption of polygenic scores [25–27]. Polygenic scores are composite weighted or unweighted aggregated locus indexes (e.g., a set of SNPs) that assess prominent loci of a given gene or identified as risk loci in GWA studies. For example, The Schizophrenia Working Group of the Psychiatrics Genetic Consortium [28] reported on a major GWA study of 36,989 cases and 113,075 controls in which they identified 108 schizophrenia-associated genetic loci from an evaluation of more than nine million variants. They also derived a weighted polygenic liability score that was based on risk profile scores (RPS) from a previous discovery GWA schizophrenia study [26]. The RPS for the larger confirmation sample increased its overall strength of association relative to the discovery sample and accounted for about 7 percent of the variance on a schizophrenia liability scale. About one-half of the 7 percent of variance accounted for (3.4 percent) was attributable to loci that met genome-wide significance. As another example, in a GWA study reported by Speliotes et al. [29] with almost 250,000 subjects, 40 genetic loci were identified that were significantly associated with the phenotype of height. However, the percentage of variation in height accounted for by these 40 loci was approximately 5 percent; note that the estimated heritability for height is about 80 percent. Based on these findings, Goldstein [30] estimated that it would require the identification of 93,000 SNPs to account for 80 percent of the population variance in height. Polygenic scores ranging from several SNPs to thousands of SNPs have been derived and associated with phenotypes such as cocaine dependence [31], alcohol problems [32], behavioral disinhibition [33], ADHD [34], intelligence [35], sensation seeking [36], Body Mass Index and obesity [29, 37], smoking cigarettes [38, 39], smoking cessation [12], reward deficiency syndrome [10], and prostate cancer [40]. Small effect size estimates (< 5 percent with many < 1 percent) characterize these polygenic scores with regard to their prediction of respective phenotypes and this remains a significant concern even with the presumed more comprehensive and higher statistically powered polygenic scores.

Conceptually, polygenic scores are accommodating to the notion of equifinality, that is, different gene combinations may produce the same end state (e.g., multiple pathways to the same disease or trait), as well as multifinality (or pleitropy), that is single genes, or derived polygenic scores, may influences a range of seemingly independent phenotypes. An interesting example of genetic pleiotropy was reported by Andreassen et al. [41] in the study of common gene variants that influenced both multiple sclerosis and schizophrenia, but not bipolar disorder. Common loci for schizophrenia and multiple sclerosis were associated with the major histocompatibility complex (MHC), though the associated human leukocyte alleles (HLA) were related in the opposite direction for the two phenotypes. That is, alleles associated with an increased risk for multiple sclerosis were associated with a decreased risk for schizophrenia.

Polygenic scores may also be useful in identifying both common (cross-disorder) polymorphisms as well as polymorphisms associated with specific disorders [42, 43]. For example, several studies have identified genes (e.g., MAO-A, GABRA2, and CHRM2) associated with a broader dimension of behavioral disinhibition or undercontrol that includes phenotypes such as alcohol, tobacco, and substance use, childhood conduct problems, and antisocial personality disorder [21, 33, 44]. Presumably, the common liability may impact underlying biological, cognitive, and affective functioning through a range of common mechanisms (e.g., brain reward center; stress reactivity or regulatory system) to influence co-occurring disorders (e.g., alcohol dependence, antisocial personality disorder). However, in addition to a common liability across these disorders, there has also been evidence of gene-phenotype specificity. For example, a consistently identified alcohol-disorder specific gene has been ALDH*2, which is associated with the alcohol metabolizing enzyme of *aldehyde dehydrogenase*. A mutation of the ALDH*2 gene on chromosome 12 yields an enzyme that cannot convert and eliminate acetaldehyde at the usual levels and rate. At high levels, acetaldehyde is toxic and induces a range of unpleasant responses (e.g., "facial flushing," nausea, heart palpitations). Hence, the study of cross-disorder or cross-trait relationships may be enhanced by polygenic scores that include both common (aggregated) and unique (disaggregated) sets of genes and environments.

Statistically, polygenic scores may be advantageous in that, similar to two-stage GWAS procedures, polygenic scores may reduce the substantial burden of multiple comparison adjustments from testing a potentially substantially larger number of SNPs to a limited few. This is especially important for many behavioral studies (e.g., small-to-midscale size epidemiologic surveys, brain imaging studies, pharmacologic studies, treatment studies, preventive intervention studies) where sample sizes are likely to continue to fall far short of sample sizes used in GWAS or large-scale NGS studies. Two-stage procedures consider reducing the number of SNPs to test for GE interactions from millions to perhaps a few hundred; in many behavioral studies the effort will be to reduce the number of SNPs or polygenic indexes from millions to a maximum in the single, low double digits. As the genomic and GE literature progresses, it will have the opportunity and challenge to provide refinements and guidelines for larger scale polygenic scores that are confirmed across samples, as well as network analyses, to reveal a better system of relationships to target in more circumscribed GE studies of complex phenotypes.

Measurement Considerations for Polygenic Scores

There are (at least) two general issues that merit consideration when deriving polygenic scores—the first is the issue of weighting and the second is the issue of measurement model representation.

Consistent with the historical precedents in epidemiology and psychometric theory of deriving composite scores from a set of indicators or items, there are a range of ways to "weight" the derived index. One method is simply to unit weight each allele by assigning a score of 0, 1, or 2 corresponding to directionality of risk (if known) or to the most frequent homozygous pair being assigned a score of 0, the heterozygous pair a score of 1, and the minor allele frequency pair being assigned a score of 2. Alternatively, 0 and 1 scores are sometimes assigned if one homogenous and the heterogeneous allele are combined to form a risk or protective allele group. Unit weighted scores are then summed to form the polygenic score. Different approaches have been used to derive "weighted" (instead of unit or unweighted) polygenic composite scores. For example, Speilotes et al. [29] weighted SNPs for BMI based on univariate regression-weighted scores derived from a large sample (n = 249,796) and these weights were used as a referent in other studies to weight SNPs predicting BMI [37]. Others have identified clusters of SNPs based on the magnitude of p-values cutoffs (e.g., $p < 0.20$, $p < 0.15$, $p < 0.10$, $p < 0.05$, $p < .01$) and evaluated the strength of associations with the targeted phenotype based on distinct clusters [26]. Still others with family data derived principal component scores that included unique family clustering relationships to weight SNPs [45]. New methods of forming polygenic scores are likely to be derived in the near future that integrate both statistical and biological functional information for specific systems and subsystems that impact single and cross-disorder/cross-trait phenotypes.

Another issue related to polygenic indexes is the role of measurement models and differences in measurement approaches. A common approach to develop polygenic scores is to follow traditional psychometric theory and use various models (principal component and common factor models) to address issues of dimensionality and scaling weights. This approach typically assumes an underlying latent trait or traits with items (SNPs) that are more highly correlated within than between traits. The overall magnitude of the correlations among items or indicators and sample size determine the internal consistency of the dimension. For example, Nock et al. [46]

used seven gene-based latent variables (or factors), each with three SNPs as manifest indicators, in addition to latent variables measured by other indicators (e.g., blood pressure by systolic and diastolic blood pressure; lipids by high-density lipoproteins and triglycerides) to measure metabolic syndrome that was then associated with coronary heart disease. By contrast, other summative models (whether weighted or unweighted) do not include an error component and high correlations among indicators are typically not highly desired because redundancy (high linkage disequilibrium) among SNPs does not contribute unique variance in accounting for the phenotype of interest. In some two-stage procedures, a first stage is used to prune those SNPs in high linkage disequilibrium. Prescreening of SNPs for linkage disequilibrium (e.g., if LD > 0.5) is often employed to "prune" the set of SNPs to be used in subsequent analyses [47, 48], though some note no differences in findings whether the SNP data are pruned or not [26]. Given that the development of polygenic scores is still in the early stages of development, it will be important to consider optimal measurement models and weighting procedures to develop and refine scores that map well on to the phenotypes of interest.

Challenges for Polygenic Scores

In addition to potential strengths and advantages of utilizing polygenic scores, there are also a number of challenges ahead in developing and adopting such indexes. For example, a limitation of polygenic scores is that they may mask biological functionality and the identification of the underlying mechanisms (e.g., network of relationships). That is, the aggregated scores may be limited in identifying which specific SNPs are impacting what underlying metabolic or neurobiological systems. There are also issues with identifying and labeling alternative polygenic scores. It would be possible to replicate findings from a similarly named polygenic score across studies, but nevertheless have a different set of SNPs within the score varying across studies. Conversely, a similar set of SNPs could be derived for scores that are labeled differently. For example, there is considerable overlap in the SNP set for the Genetic Addiction Risk Score [10] and the dopamine gene polygenic index created by Derringer et al. [31] that were associated with dependence symptoms among cocaine users. This may lead to enumeration complexity in that an exceedingly large number of polygenic scores could be derived unless systematic efforts are made to organize such scores in a database for larger scientific and public consumption. Furthermore, the formation of polygenic scores will not necessarily facilitate the identification of causal variants, though ideally it will increase the probability of such identification by reducing the possible set of potential SNPs to be investigated via alternative methods that can focus on biological mechanisms (e.g., gene expression data, knockouts).

Another limitation of polygenic scores is that there will be differences among investigators in the development of research strategies and algorithms to select an optimal set of genotypes and SNPs for inclusion. For example, is it better to optimize at the gene level (e.g., genetic loci of dopamine genes), with reference to specific biologic systems (drug metabolism; brain reward center), with regard to a specific phenotype (e.g., stress), or with regard to an optimal set of empirically/statistically based GE optimization methods? In essence, what are the guidelines for selecting the genetic loci to form a polygenic score both to optimize prediction and to provide a guide to biological mechanisms? Another issue that needs to be addressed is how should dependent variables be selected in a manner so as not to generate different polygenic scores for every phenotype, or potentially the same phenotypes in different studies? That is, a different optimized polygenic score could be derived for the same person for alternative substance use phenotypes such as onset, frequency, quantity, adverse consequences, abuse, dependence, cessation, and relapse. More efficient polygenic scores are needed to address common and unique features of development and disease changes (e.g., progression, termination). Finally, for the study of GE interactions, it may be more difficult to extract potential explanations of mechanisms if biologically the polygenic scores are related to multiple different functional systems and mechanisms that may account for the obtained polygenic score × environment interaction.

Limitations of Using GWAS Findings to Guide the Development of Polygenic Scores for the Study of G × E Interactions

In addition to the more general challenges of polygenic scores described previously, there are also specific challenges to deriving polygenic scores based solely on GWAS findings to study polygenic score × E interactions. For example, multiplicative G × E interactions are, by definition, expressions of conditional relationships among variables rather than "main effects." Genome-wide searches that identify main effects may not yield the optimal set of SNPs or loci that likely interact with environmental factors to predict the specific phenotype of interest. That is, various discovery and confirmation samples may be used in large-scale replication GWA studies to identify a core set of SNPs for a given phenotype, but there is nothing inherent in this core set that necessitates that they are optimal for studying $G \times E$ interactions. An example of this was illustrated in an analysis of relationships between cigarettes smoked per day and 91 prominent SNPs identified from a meta-analysis [39]. The polygenic score for the 91 SNPs was not statistically significant for the full sample; however, a polygenic score was significantly associated with cigarettes smoked per day at ages 20 and 24 but not ages 14 and 17. Hence, the developmental interaction (polygenic score by time/age) was masked by the pooled sample across ages and the assumption was incorrect that the relationship between the polygenic score and smoking was not conditional on age.

Another limitation of using "main effects" GWAS-derived gene and polygenic scores relates to the sensitivity of the discovery and confirmation analytic procedures to detect the kinds of crossover interactions that typify some $G \times E$ relationships, such as those suggested by the differential susceptibility hypothesis [49, 50]. The differential susceptibility hypothesis suggests that some genes manifest plasticity rather than only risk or protectiveness, and that the nature or direction of the significant $G \times E$ interaction is contingent on the environment. Hence, a plasticity gene may be simultaneously associated with a higher risk of poor outcomes under certain environmental conditions (low SES, impoverished neighborhood) but associated with higher-quality outcomes under other environmental conditions (high SES, enriched neighborhood). That is, individuals vary not only with regard to how much they are negatively impacted by stressful environments (e.g., low-cohesion neighborhoods, parental neglect, parental disorders), but also how much they are positively impacted by enhanced or promotive environmental factors (e.g., strong support from teachers, resources such as playgrounds in the neighborhood) [49, 51]. Statistical tests for $G \times E$ interactions associated with the differential susceptibility hypothesis evaluate regions of significance at both the upper and lower end of the distribution [52]. The concern with using main effects GWAS-derived gene and polygenic scores is that the form of the modeled interaction effects (at both the lower and higher end of distribution) may not be sensitive to these crossover effects.

Measurement of the Environment

In order to maximally utilize polygenic scores to test G × E interactions, another challenge is to select and measure appropriate aspects of the environment with sufficient sensitivity and variability to facilitate hypothesis testing [53]. For many complex phenotypes, it is recognized that there are multiple-level environmental influences (e.g., individual level such as stressful events, family and peer factors, neighborhood factors, cultural factors) that may singly or jointly significantly interact with genomic factors to impact the occurrence, timing of the onset (e.g., early vs. later onset), severity, and comorbidity of various disorders [54, 55]. Furthermore, because environments are dynamic (i.e., changing and emerging across time) and sometimes specific to a particular disorder or behavioral phenotype, there is not a finite list of environments that can be easily enumerated for use across complex phenotypes. Nevertheless, there are already (at least) four environmental domains or areas that are now briefly discussed that merit consideration in future G × E studies, and others that are all likely to emerge with additional G × E studies.

First, there has been substantial research supporting the role of stressful life events in general, and childhood maltreatment in particular, that have yielded reasonably consistent findings for G × E associations between the serotonin transporter 5-HTTLPR and stress/childhood maltreatment on adverse adult outcomes such as depression [56] or PTSD [57]. Second, many studies have recognized the role of social environments (e.g., parents, peers) that may significantly interact with specific genes (e.g., DRD4, GABA) to impact childhood and adolescent internalizing and externalizing problems, including substance misuse [58–60]. Third, new technologies and statistical methods in environmental health (e.g., sensors for environmental pollutants, refined neighborhood spatial measures and modeling) may provide valuable tools to further assess gene and polygene by environment interactions for conditions such as asthma and other respiratory conditions as well as cancer [61].

Fourth, the role of repeated measures data to address developmental issues has yielded some initial interesting findings. For example, several studies have reported significant G × E interactions for parenting characteristics (e.g., parental monitoring, parental support) on child and adolescent behavioral problems and substance use [59, 62, 63]. Likewise, significant G × E interactions for deviant or substance using peers on adolescent and adult substance use have been reported [59, 60, 64]. Some of these findings suggest that deviant peer environmental influences are stronger in adolescence but that genetic influences become more prominent in adulthood [59, 60]. Longitudinal findings by Uhl et al. [12] using developmental trajectory analyses and a polygenic score identified a subgroup of young adults characterized both by a higher rate of smoking cessation and a lower use of all substances during adolescence. Hence, the study of what genes interact with time (age) to predict specific complex phenotypes of interest (e.g., substance use, externalizing problems, smoking cessation propensity) provides a powerful and underutilized paradigm for investigating G × E relationships. Likewise, as noted by Bakermans-Kranenbourg and IJzendoorn [49] and others [62, 63], intervention designs provide a very useful and powerful methodological approach for studying G × E relationships because randomization of participants to groups reduces the potential of G × E correlations as an alternative explanation if G × E interactions are obtained.

Proposed Second-Generation Candidate Gene Approach

Advances in methodology and technology often precipitate modifications in approaches to knowledge generation. For example, limitations in GWAS has spawned NGS methods such as whole genome sequencing that includes both a more extensive and intensive assessment of genomic material (e.g., deep sequencing) to capture information beyond that provided in the initial GWAS-based sequencing studies [27, 65]. Included in whole genome sequencing are not only the measurement of common genetic variants, with minor allele frequencies ≥ 5 percent, but also other genomic sites such as rare variants (with minor allele frequencies < 5 percent) and other indels (insertions and deletions), chromosomal rearrangements, and copy number variants (CNVs).

Given the identified limitations of initial stage candidate gene studies and candidate gene by environment studies [24, 66, 67], it would be beneficial to reconceptualize and apply principles in what may be described as second-generation candidate gene and candidate gene by environment

Table 10.1
First- and second-generation candidate GE studies

First-Generation Candidate G × E Studies	Second-Generation Candidate G × E Studies
1. Single SNP or VNTR	1. Multiple SNPs
2. Assumed large effect size per SNP	2. Assumes smaller effect size per SNP
3. Assumed knowledge of key SNPs/VNTRs	3. Less certain of key SNPs/VNTRs
4. Not influenced by GWA studies	4. More influenced by GWA studies
5. Statistical corrections typically not applied	5. Statistical corrections more often applied
6. Replication rare	6. Replication increasing
7. Candidate gene method only	7. Joint use of candidate gene with other methods (e.g., gene expression data)

interaction studies. Table 10.1 summarizes some of the distinctions that might be used to describe them.

A key motivation for a second-generation candidate gene approach is, as described previously, that many behavioral, intervention, neuroimaging, and pharmacogenetic studies of GE relations for complex phenotypes will not be powered at the same level as GWAS or NGS studies. This need not serve as a deterrent to pursuing G × E investigations, but nevertheless such studies must address prior methodological weaknesses to the extent possible and use available resources (e.g., GWA and NGS study findings, prior theories and research, findings from infrahuman and human studies) and a consilience research orientation [68, 69] to guide future research endeavors.

In distinguishing first- and second-generation candidate GE studies, it is important to move beyond single SNPs or VNTRs to multiple SNPs, VNTRs, or other loci (e.g., rare variants). It is not that single, common variant candidate gene approaches cannot still be of value to the literature, only that they are not likely to account for large portions of variance and, of course, they might not be the causal variant of interest for the phenotype and may not replicate across samples. The selection of multiple SNPs (or other loci) should ideally be driven by prior research, such as polygenic indexes derived and validated in larger GWAS or NGS studies, or consistency of experimental and correlational findings from infrahuman and human studies. More limited sample sizes in behavioral science studies of complex phenotypes will place restrictions on the number of hypotheses that can be feasibly tested; prior empirical findings (e.g., via meta-analyses, GWAS, and gene network analyses) may assist in serving as fountains of knowledge to guide reductions in genomic loci and environments selected for study.

Many first-generation gene and GE studies were based on the common variant model, and one assumption of this model was that there would likely be moderate to large effect sizes per SNP or VNTR. Hence, although it was recognized that for complex phenotypes many polymorphisms may be required to account for large portions of variance that would approach extant heritability estimates, the standard thinking was not that this number would be on the order of potentially thousands of SNPs but rather something more modest like 20–30 (or even hundreds of) SNPs. In second-generation studies, it would be beneficial to recognize that for most complex phenotypes, any single, common variant locus (e.g., SNP, VNTR) is likely to yield small effect sizes, and therefore sample size considerations must assume a greater weight both in planning and interpreting GE effects. Note that rare variants may yield larger effect sizes, but much research remains to be completed on rare variants for complex phenotypes before any major conclusions can be drawn about their explanatory power and therapeutic value for a broad range of complex phenotypes.

Related to the first-generation notion that effect sizes were likely to be moderate to large, many first-generation candidate gene studies also assumed baseline knowledge about a given genetic substrate (i.e., VNTR, SNP) and associated neurotransmitters and promoters that had been identified in the fields of psychiatry and infrahuman studies [67]. The general assumption was that genetic loci associated with these neurotransmitters would be valuable initial targets to explore candidate genes and candidate G × E interactions, and that these targets would yield moderate to large effect sizes and therefore adequate power was not pursued as rigorously as is needed. Second-generation studies need to be more sensitive to probable small effect sizes for individual loci and need to further draw upon other models and sources of data (e.g., rare variants, network analyses, biosystems approaches) that attempt to map larger units of development and regulation rather than a single loci or mutation that is impacting the phenotype of interest.

Similarly, first-generation studies were not overly influenced by GWA studies in terms of the selection of genetic loci. Part of the reason for this was timing (i.e., historically, many candidate gene studies preceded GWA studies) and part of this was that many of the GWAS findings were neither replicating nor providing consistent loci upon which to follow up more intensively [23, 70, 71]. Nevertheless, in second-generation candidate GE studies, findings from GWAS may be one source used to help identify specific genes (to be examined at the gene level) and polygenic indexes to help guide the selection of the G component of the G × E interaction. Of course, while the use of GWAS findings are valuable to include in considering the investigation of G × E relationships, they are only one source, because they do not provide guidance as to which E components should be considered or which G × E interactions because analyses of traditional GWA studies do not provide information on which gene loci might interact with which Es to impact the expression of a given phenotype. Nevertheless, it is important to consider the findings on main effects from GWA, NGS, and meta-analytic studies as potentially useful resources for second-generation candidate G × E studies.

First-generation candidate G × E studies infrequently applied statistical corrections for multiple testing. It would be beneficial for second-generation candidate G × E studies to appropriately apply statistical corrections. GWA studies have developed valuable approaches (e.g., FDR; permutation testing) to apply to multiple test data to yield conservative, but not overly conservative, statistical corrections. Similarly, replication was less frequently completed in first-generation studies, and this needs to occupy a more central role in second-generation candidate G × E studies. A central repository or organizing website might be of value for behavioral science investigators to facilitate such G × E replication studies. For some areas of investigation (e.g., longitudinal research designs, RCTs), replication studies remain a challenge because of differences in major design parameters (e.g., sampling designs, measures, intervals between measurement). These areas will require additional thought as to how best to address the replication issue.

Finally, first-generation studies relied exclusively on the candidate gene approach; second-generation studies may use more than one method (e.g., candidate gene, gene expression, and/or DNA methylation profiling methods) to pursue in more detail both associations and biological functionality of targeted candidate genes. For instance, Guintivano et al. [72] used genome-wide DNA methylation profiling with postmortem brains and peripheral blood from three living groups to derive gene expression data to identify an additive epigenetic and genetic association with suicide at rs7208505, which is in the 3′ untranslated region of the SKA2 gene. Further analyses indicated that SKA2 gene expression was lower in suicide dece-dents and was possibly mediated by an interaction between genetic and epigenetic variation in rs7208505 and an intronic microRNA (miR-301a). Thus, the use of more than one method may be a useful strategy in some applications to further examine the biological functionality of candidate genes.

In summary, second-generation candidate gene and candidate G × E studies may draw upon limitations of both first-generation candidate G and G × E studies and GWA studies to guide more rigorous testing of prob-able gene-environment relationships. Consilience, which involves draw-ing upon common and unique findings across different fields of study and different species, is likely to be an organizing vehicle, as well as drawing upon the theoretical and empirical literature, including GWA and meta-analytic studies, to guide multilocus or polygenic G × E models of pheno-types related to intermediate mechanisms and outcomes.

Research Questions and Conceptual Models of G × E Relationships

In considering the kinds of conceptual and statistical models needed to address G × E relationships, it is imperative to consider what research ques-tions such models seek to explain, and of course these may vary contin-gent on the phenotype of interest. Within the area of substance use among youth, some of these questions are:

• Which G and E factors contribute to some youth initiating use and mis-use earlier than other youth?
• Which G and E factors contribute to the escalation and maintenance of high levels of substance use for youth?
• Which G and E factors contribute to youth decreasing their use of sub-stances after having engaged in high levels of use?
• Which G and E factors contribute to substance abuse disorders?
• Are G and E factors similarly influential in adolescence and adulthood?
• What kinds of interventions (E) work best in conjunction with which genes and what kinds of substance users/abusers to maximize effectiveness?

Model Representations of Current and Future GE models

In seeking to address the research questions posed above, as well as other more dynamic relationship research questions (e.g., multiple factor change models), it is of value to examine where the field has been with regard to the study of GE relationships for complex phenotypes, as well as directions of where it might be going. This will be accomplished via a series of path diagrams to illustrate how conceptual and statistical models are evolving in the study of GE relations. In this section, aggregated polygenic by environment models are distinguished from disaggregated polygenic by environment models. Aggregated models use the composite polygenic index score in the GE models whereas the disaggregated models may use common and unique SNP sets to predict different phenotypes in a multivariate context.

Aggregated Polygenic × Environment Models

The model provided in figure 10.1 illustrates initial stage research in the field in that a single locus (e.g., VNTR or SNP) was investigated for association with a given phenotype, in this instance the relationship between DRD4 and alcohol dependence. Although suitable rationales have been provided for investigating such a relationship [10, 31], findings from models of this sort for complex phenotypes typically tend either not to replicate or to manifest small effect sizes (i.e., explain a small percentage of variance in the complex phenotype of interest). Furthermore, there has been an increasing recognition in the literature that identified candidate loci may be associated with but not be the causal variant underlying the disease or trait of interest. The model provided in figure 10.2 provides an extension of figure 10.1 in that now a polygenic score is used rather than a single locus score to predict (or be associated with) the phenotype. As described previously, there are a number of motivations for using polygenic scores,

Figure 10.1
Single gene predicts alcohol dependence model.

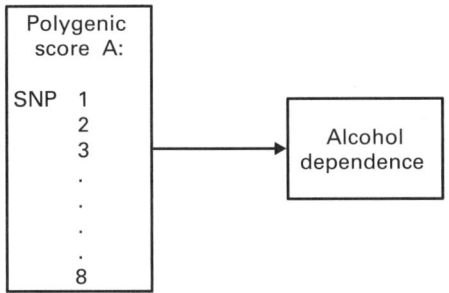

Figure 10.2
Polygenic score A predicts alcohol dependence model.

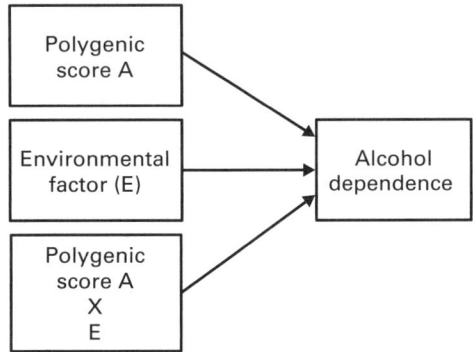

Figure 10.3
Polygenic score A × environmental interaction predicts alcohol dependence model.

including the attempt to increase effect size estimates and to reduce the overall number of statistical tests conducted. Hence, figure 10.2 provides a more enriched, or at a minimum a more flexible, model to investigate associations between multiple loci and a complex phenotype. As described previously in this chapter, polygenic scores of this sort have been used to predict a broad range of phenotypes.

The model portrayed in figure 10.3 is expanded to include not only a polygenic score but also an environmental factor (E) and a polygenic score by E interaction. There have been few research applications of this sort, although Salvatore et al. [32] reported significant interactions between two separate environmental factors (peer deviance, parental knowledge of child's behaviors) and a polygenic risk score on number of youth alcohol

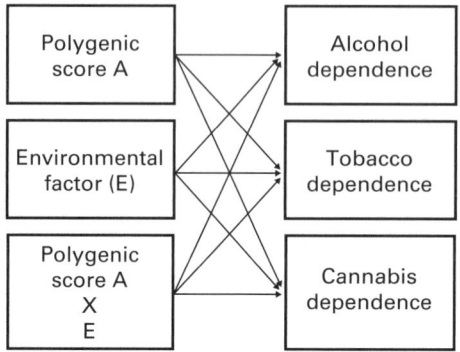

Figure 10.4
Polygenic score A × environmental interaction predicts multiple substance use disorders.

problems. The statistically significant interactions indicated that the polygenic score was more influential (i.e., associated with more youth alcohol problems) when parental knowledge was low or peer deviance was high. This research application provides a clear illustration of how models such as the one represented in figure 10.3 can be used to incorporate both polygenic scores and features of the environment to evaluate their joint contributions via multiplicative interactions on outcomes of interest. It further illustrates how utilizing information from prior empirical research both on selected SNPs and two environmental factors can be used to test hypothesized relationships.

Figure 10.4 provides a model that incorporates an additional step in the progression of models presented in that it also incorporates multiple outcome variables rather than a single outcome variable to evaluate the influences of a polygenic score, an environmental factor, and the polygenic score × E interaction. As described in the previous section on significant research questions in substance abuse and other areas of study (e.g., cross-disorders), multivariate outcomes can be quite beneficial in identifying common and unique relationships between polygenic × E relationships. The model in figure 10.4 could be used to examine the similarities versus differences in the polygenic score by E relationships regarding three phenotypes that are typically assumed to share some common sources of genetic variation [21, 73]. Given prior research on the polygenic score or the E factor, it would be possible to test, for example, if the polygenic score significantly influenced all three phenotypes, but the polygenic × E factor influenced only one of the phenotypes. A constrained structural equation modeling approach may be highly suitable to addressing hypotheses about presumed common and unique relationships among polygenic and E factors.

Disaggregated Polygenic Scores × Environment Models

Disaggregated polygenic scores by E models enable one to address a variety of research questions such as common and unique SNPs associated with general and specific phenotypes (e.g., general disinhibition and sensation seeking) or longitudinal models where different SNPs may predict behavioral phenotypes at differ points in the lifespan. The model portrayed in figure 10.5 provides a representation of how disaggregated SNPs may be useful in mapping onto both common and unique elements of a multivariate outcome domain. A factor of general disinhibition or behavioral undercontrol has commonly been identified to underlie a broad range of behaviors ranging from childhood behavior disorders, substance abuse disorders, impulsivity, and antisocial personality disorder [44, 74, 75]. (See also Arcos-Burgos et al. [76] for a common network model underlying these multiple phenotypes.) However, in addition to a shared vulnerability, there is also trait-specific variance associated with the indicators of the broader dimension such that there is both reliable shared variance with the broader general disinhibition dimension and trait- or construct-specific variance.

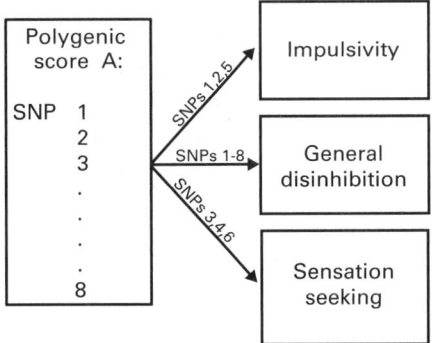

Figure 10.5
Polygenic score A predicts common (cross-phenotype) and unique (specific) components of disinhibition.

The model in figure 10.5 provides a method of determining which specific SNPs impact the general disinhibition factor and which SNPs influence the trait- or construct-specific variance. In the model representation, all eight SNPs are proposed to be associated with general disinhibition, but only SNPs 1, 2, and 5 are proposed to be associated with impulsivity, and SNPs 3, 4, and 6 are proposed to be associated with sensation seeking. Hence, this representation provides a method of testing the convergent-discriminant validity of sets of SNPs in relation both to a common vulnerability and to specific unique elements.

The model represented in figure 10.6 is expanded in two ways. First, two polygenic scores are used rather than one, and second, the multivariate outcome is now one based on longitudinal assessments related to life-course trajectories of adult alcohol dependence with antecedents in childhood (e.g., childhood externalizing problems) and adolescence (adolescent alcohol use). Polygenic indexes are beginning to emerge in the literature, and across time it is likely that refinements will contribute to reliable and valid scores that are associated with specific outcomes. However, to increase

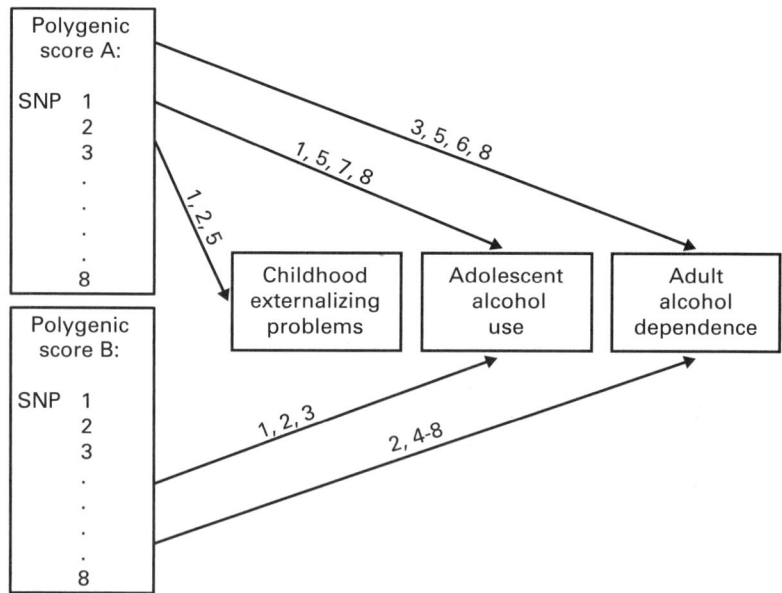

Figure 10.6
Two polygenic scores predicting time sequence to alcohol dependence.

effect sizes for phenotypes, it is probable in many areas of study that more than one polygenic index will be utilized. This is especially the case for cross-disorder or cross-trait analyses, or for subtypes that either currently exist or will emerge across time. For instance, in the alcohol studies literature it is common to discuss both a behavioral undercontrol (externalizing) pathway and an internalizing (alcohol with co-occurring depression or anxiety) pathway [55, 77]. To optimize prediction of life course alcohol sequence data, it may be beneficial to include polygenic indexes derived for both externalizing (or behavioral undercontrol) and internalizing problems. The illustration in figure 10.6 suggests, for example, that SNPs 1, 2, and 5 from polygenic score A are associated with childhood externalizing problems; adolescent alcohol use is associated with SNPs 1, 5, 7, and 8 from polygenic score A and SNPs 1, 2, and 3 from polygenic score B; and adult alcohol dependence is associated with SNPs 3, 5, 6, and 8 from polygenic score A and SNPs 2 and 4–8 from polygenic score B. In sum, this model facilitates the testing common and unique relationships between selected SNPs of two major polygenic indexes and phenotypes that reflect the life course development of alcohol dependence.

Figure 10.7 is an expansion of the representation in figure 10.6 in that multiple (two in this instance) environmental factors are added to the two polygenic index model, and polygenic by E interactions are included. Although challenging, this representation could be used to model statistically with greater precision the relationships between common and specific sets of SNPs and common and specific environmental factors as they impact the time course of a given disease or disorder (adult alcohol dependence in the example). Although challenging to contemplate and implement, models such as the one presented in figure 10.7 are likely to be necessary for complex phenotypes to increase effect size estimates, to understand the unfolding $G \times E$ causal processes, to translate findings into clinical practice, and to move toward personalized medicine.

Structural Equation Models (SEMs) of G × E Relationships

In the previous section a sequence of aggregated and disaggregated polygenic × environment models was described via path diagrams to provide one possible roadmap of how the study of $G \times E$ relationships may progress in the future. As the diagrams illustrated, as we move from bivariate and

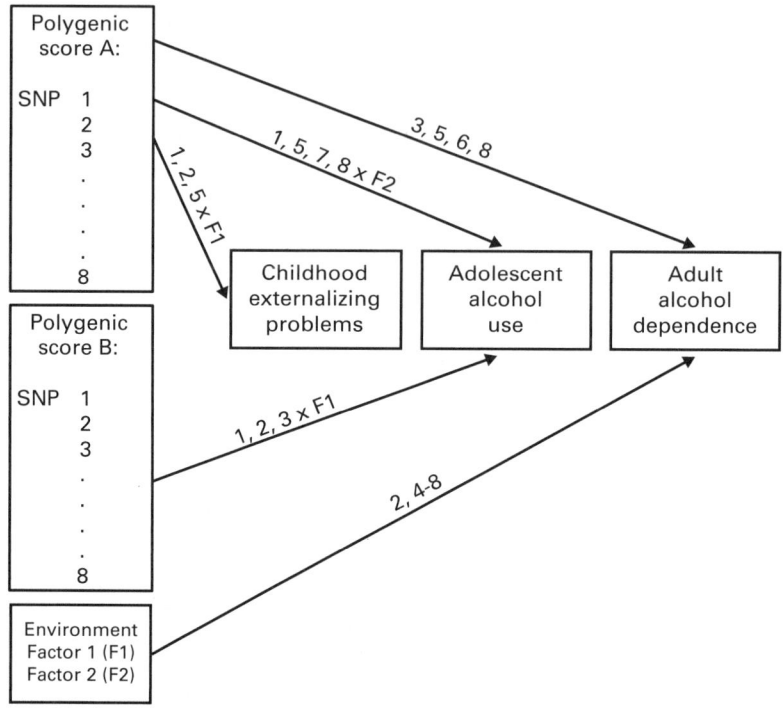

Figure 10.7
Two polygenic scores and two environmental factors predicting alcohol dependence.

cross-sectional research designs with one dependent variable to the multi-variate longitudinal research of G × E relationships, the complexity of the associated statistical models and associated equations increases. Guided by theory, conceptual models, and prior empirical findings (e.g., from GWAS or initial candidate gene study findings that bear replication and extension), univariate logistic or linear regression models may be limited in their capability to address some important research questions related to G × E relationships for complex phenotypes. Although used widely in the broader social and behavioral sciences, with some notable exceptions [46, 78, 79], SEMs have had limited applications in the study of G × E relationships.

There are a number of reasons to consider SEM methodology as a valuable approach in the investigation of G × E relations with complex phenotypes. First, at a general level of modeling, SEMs represent a flexible approach that has the capacity to model the multiple SNPs or other genetic loci, multiple environmental factors, multiple occasions of measurement, and multiple outcomes such as those presented in figures 10.6–10.7. As such, it serves as a useful conceptual as well as an analytical tool for envisioning presumed causal processes involving $G \times E$ mechanisms across time. Second, statistical power may be increased in multiple ways such as increasing reliability via the use of latent variables for predictors, mediators, and moderators, or using covariates to reduce error variance of the dependent variable(s), thereby increasing power. Third, SEM is able to accommodate alternative research designs (e.g., case-control, case only, RCT, observational) to facilitate the study of $G \times E$ hypotheses.

Summary

Complex phenotypes that may include $G \times E$ and $G \times G$ interactions terms to explicate causal processes pose multiple conceptual and methodological challenges to the field. A major knowledge gain from GWA studies has been that the common variant, common cause model of genetic etiology has fallen short of initial expectations; effect sizes have been substantially smaller than anticipated, and the number of loci needed to account for even a small percentage of variance in the phenotype of interest appears to be quite large. However, the field has been adaptive and is developing alternative ways to address the new conceptual and methodological issues that have arisen and will continue to arise. In this chapter, a number of statistical approaches to address some of these challenges have been briefly described (e.g., a focus on the gene level rather than SNP level; two-stage approaches), including the potential value and limitations of polygenic scores. The chapter also suggests the need of a second-generation candidate gene approach to adapt to first-generation limitations and to strengthen efforts to study $G \times E$ interactions. The sample sizes associated with many phenotypes and areas of study in the literature (e.g., clinical trials, neuroimaging studies) are highly unlikely to yield sample sizes in the area of GWA and NGS studies (e.g., 200,000–300,000 participants). However, by building upon prior limitations in the candidate gene literature, using findings from GWA and NGS studies and meta-analyses, using multiple methods of analyses (e.g., gene expression analysis; methylation analysis), and using theory and prior substantive research to guide hypothesis testing, progress can be made on $G \times E$ interactions for complex phenotypes. Several illustrative path models were provided in this chapter to provide a visual frame for how we have approached G × E interactions in the past, and how, going forward, we might proceed to investigate multiple polygenic by multiple environmental models. This level of complexity may be necessary to advance the field to address the many exciting research questions of interest, as well as the challenges that confront us as we attempt to move this knowledge from discovery to practice.

Acknowledgments

This research was supported by NIDA Grant Number DA027827 and NIAAA Grant Number K05AA021143. The contents are solely the responsibility of the author and do not necessarily represent the official views of the National Institutes of Health.

References

1. Smith, P. G., & Day, N. E. (1984). The design of case-control studies: The influence of confounding and interaction effects. *International Journal of Epidemiology*, *13*, 356–365.

2. Thomas, D. C. (2010). Methods for investigating gene-environment interactions in candidate pathway and genome-wide association studies. *Annual Review of Public Health*, *31*, 21–36.

3. Benjamini, Y., Drai, D., Elmer, G., Kafkafi, N., & Golani, I. (2001). Controlling the false discovery rate in behavior genetic research. *Biobehavioral Brain Research*, *125*, 279–284.

4. Nichols, T. E., & Holmes, A. P. (2002). Nonparametric permutation tests for functional neuroimaging: A primer with examples. *Human Brain Mapping*, *15*, 1–25.

5. North, K. E., & Martin, L. J. (2008). The importance of gene-environment interaction. *Sociological Methods & Research*, *37*, 164–200.

6. Kooperberg, C., Dai, J. Y., & Hsu, L. (2016). Two-stage procedures for the identification of gene × environment and gene × gene interactions in genome-wide association studies. In M. Windle (Ed.), *Statistical Approaches to Gene × Environment Interactions for Complex Phenotypes* (pp. 11–34). Cambridge, MA: MIT Press.

7. Guo, X., Liu, Z., Wang, X., & Zhang, H. (2013). Genetic association tests for multiple traits at gene level. *Genetic Epidemiology*, *37*, 122–129.

8. Wang, T. (2016). A gene-based approach for testing gene × gene and gene × environment interactions. In M. Windle (Ed.), *Statistical Approaches to Gene × Environment Interactions for Complex Phenotypes* (pp. 71–94). Cambridge, MA: MIT Press.

9. Bierut, L. J., Agrawal, A., Bucholz, K. K., Doheny, K. F., Laurie, C., Pugh, E., et al. (2010). A genome-wide association study of alcohol dependence. *PANAS, 107*, 5082–5087.

10. Blum K., Oscar-Berman, M., Demetrovics, Z., Barh, D., & Gold, M.S. (2014). Genetic Addiction Risk Score (GARS): Molecular neurogenetic evidence for predisposition to reward deficiency syndrome (RDS). *Molecular Neurobiology*, s12035–014–8726–5.

11. Brody, G. H., Chen, Y., & Beach, S. R. H. (2013). Differential susceptibility to prevention: GABAergic, dopaminergic, and multilocus effects. *Journal of Child Psychology and Psychiatry, and Allied Disciplines, 54*, 863–871.

12. Uhl, G. R., Walther, D., Musci, R., Fisher, C., Anthony, J. C., Storr, C. L., et al. (2014). Smoking quit success genotype score predicts quit success and distinct patterns of developmental involvement with common addictive substances. *Molecular Psychiatry, 19*, 50–54.

13. Bowman, F. D. (2014). Brain imaging analysis. *Annual Review of Statistics and Its Application, 1*, 61–85.

14. MacKillop, J., & Munafo, M. R. (Eds.). (2013). *Genetic Influences on Addiction: An Intermediate Phenotype Approach*. MIT Press.

15. Hussong, A. M., Curran, P. J., & Bauer, D. J. (2013). Integrative data analysis in clinical psychology research. *Annual Review of Clinical Psychology, 9*, 61–89.

16. Windle, M., Kogan, S. M., Lee, S., Chen, Y. F., Lei, K. M., Brody, G., et al. (in press). Neighborhood by 5-HTTLPR interactions for substance use from ages 10–24 years using a harmonized dataset of African-American children. *Development and Psychopathology*.

17. Michaud, C. M., Murray, C. J. L, & Bloom, B. R. (2001). Burden of disease—Implications for future research. *Journal of the American Medical Association, 285*(5), 535–539.

18. Substance Abuse and Mental Health Services Administration. (2014). *Results from the 2013 National Survey on Drug Use and Health: Summary of National Findings, NSDUH Series H-48, HHS Publication No. (SMA) 14–4863*. Substance Abuse and Mental Health Services Administration.

19. Johnston, L. D., O'Malley, P. M., Bachman, J. G., & Schulenberg, J. E. (2013). *Demographic Subgroup Trends among Adolescents for Fifty-one Classes of Licit and Illicit Drugs, 1975–2012 (Monitoring the Future Occasional Paper No. 79)*. Institute for Social Research; Available http://www.monitoringthefuture.org/.

20. Kendler, K. S., Jacobson, K. C., Prescott, C. A., & Neale, M. C. (2003). Specificity of genetic and environmental risk factors for use and abuse/dependence of cannabis, cocaine, hallucinogens, sedatives, stimulants, and opiates in male twins. *American Journal of Psychiatry, 160,* 687–695.

21. Kendler, K. S., Chen, X., Dick, D., Maes, H., Gillespie, N., Neale, M. C., et al. (2012). Recent advances in the genetic epidemiology and molecular genetics of substance use disorders. *Nature Neuroscience, 15,* 181–189.

22. Hart, A. B., De Wit, H., & Palmer, A. A. (2013). Candidate gene studies of a promising intermediate phenotype: Failure to replicate. *Neuropsychopharmacology, 38,* 802–816.

23. Manuck, S. B., & McCaffery, J. M. (2014). Gene-environment interaction. *Annual Review of Psychology, 65,* 41–70.

24. Young-Wolf, K. C., Enoch, M. A., & Prescott, C. A. (2011). The influence of gene-environment interactions on alcohol consumption and alcohol use disorders: A comprehensive review. *Clinical Psychology Review, 31,* 800–816.

25. Plomin, R. (2013). Child development and molecular genetics: 14 years later. *Child Development, 84,* 104–120.

26. Purcell, S. M., Wray, N. R., Stone, J. L., Visscher, P. M., O'Donovan, M. C., Sullivan, P. F., et al. (2009). Common polygenic variation contributes to risk of schizophrenia and bipolar disorder. *Nature, 460,* 748–752.

27. Vrieze, S. I., Iacono, W. G., & McGue, M. (2012). Confluence of genes, environment, development, and behavior in a post Genome-Wide Association Study world. *Development and Psychopathology, 24,* 1195–1214.

28. Ripke, S., Neale, B. M., Corvin, A., Walters, J. T. R., Farh, K.-H., Holmans, P. A., et al. (2014). Biological insights from 108 schizophrenia-associated genetic loci. *Nature, 511,* 421–427.

29. Speliotes, E. K., Willer, C. J., Berndt, S. I., Monda, K. L., Thorleifsson, G., Jackson, A. U., et al. (2010). Association analyses of 249,796 individuals reveal 18 new loci associated with body mass index. *Nature Genetics, 42,* 937–948.

30. Goldstein, D. B. (2009). Common genetic variation and human traits. *New England Journal of Medicine, 360,* 1696–1698.

31. Derringer, J., Krueger, R. F., Dick, D. M., Aliev, F., Grucza, R. A., & Saccone, S. (2012). The aggregate effect of dopamine genes on dependence symptoms among cocaine users: Cross-validation of a candidate system scoring approach. *Behavior Genetics, 42,* 626–635.

32. Salvatore, J. E., Aliev, F., Edwards, A. C., Evans, D. M., Macleod, J., Hickman, M., et al. (2014). Polygenic scores predict alcohol problems in an independent sample and show moderation by the environment. *Genes, 5,* 330–346.

33. Vrieze, S. I., McGue, M., Miller, M. B., Hicks, B. M., & Iacono, W. G. (2013). Three mutually informative ways to understand the genetic relationships among behavioral disinhibition, alcohol use, drug use, nicotine use/dependence, and their co-occurrence: Twin biometry, GCTA, and genome-wide scoring. *Behavior Genetics, 43,* 97–107.

34. Hamshere, M. L., Langley, K., & Martin, J. (2013). High loading of polygenic risk for ADHD in children with comorbid aggression. *American Journal of Psychiatry, 170,* 909–916.

35. Benyamin, B., St. Pourcain, B., Davis, O. S., Davies, G., Hansell, N. K., Brion, M. J. A., et al. (2014). Childhood intelligence is heritable, highly polygenic and associated with FNBP1L. *Molecular Psychiatry, 19,* 253–258.

36. Derringer., J., Krueger, R. F., Dick, D. M., Saccone, S., Grucza, R. A., Agrawal, A., et al. (2010). Predicting sensation seeking from dopamine genes: A candidate-system approach. *Psychological Science, 21,* 1282–1290.

37. Belsky, D. W., Moffitt, T. E., Sugden, K., Williams, B., Houts, R., McCarthy, J., et al. (2013). Development and evaluation of a genetic risk score for obesity. *Biodemography and Social Biology, 59,* 85–100.

38. Treutlein, J., Strohmaier, J., Frank, J., Muhleisen, T. W., Degenhardt, F., Wit, S. H., et al. (2014). Smoking behavior: Investigation of thee coaction of environmental and genetic risk factors. *Psychiatric Genetics, 24,* 279–280.

39. Vrieze, S. I., McGue, M., & Iacono, W. G. (2012). The interplay of genes and adolescent development in substance use disorders: leveraging findings from GWAS meta-analyses to test developmental hypotheses about nicotine consumption. *Human Genetics, 131,* 791–801.

40. Aly, M., Wiklund, F., Xu, J., Isaacs, W. B., Eklund, M., & D'Amato, M. (2011). Polygenic risk score improves prostate cancer risk prediction: Results from the Stockholm-1 Cohort Study. *European Urology, 60,* 21–28.

41. Andreassen, O. A., Harbo, H. F., Wang, Y., Thompson, W. K., Schork, A. J., Mattingsdal, M., et al. (2014). Genetic pleiotropy between multiple sclerosis and schizophrenia but not bipolar disorder: Differential involvement of immune-related gene loci. *Molecular Psychiatry, 20,* 207–214.

42. McGrath, L. M., Weill, A. S., Robinson, E. B., Macrae, R., & Smoller, J. W. (2012). Bringing a developmental perspective to anxiety genetics. *Development and Psychopathology, 24,* 1179–1193.

43. Smoller, J. W., Kendler, K., & Craddock, N. (2013). Identification of risk loci with shared effects on five major psychiatric disorders: A genome-wide analysis. *Lancet, 381*(9875), 1371–1379.

44. Dick, D. M. (2007). Identification of genes influencing a spectrum of externalizing psychopathology. *Current Directions in Psychological Science, 16*(6), 331–335.

45. McGue, M., Zhang, Y., Miller, M. B., Basu, S., Vrieze, S., Hicks, B., et al. (2013). A genome-wide association study of behavioral disinhibition. *Behavior Genetics, 43,* 363–373.

46. Nock, N., Wang, X., Thompson, C. L., Song, Y., Baechle, D., Raska, P., et al. (2009). Defining genetic determinants of the metabolic syndrome in the Framingham Heart Study using association and structural equation modeling methods. *BMC Proceedings, 3*(Suppl 7), S50.

47. Coon, H., Piasecki, T. M., Cook, E. H., Dunn, D., Mermelstein, R. J., Weissa, R. B., et al. (2014). Association of the CHRNA4 neuronal nicotinic receptor subunit gene with frequency of binge drinking in young adults. *Alcoholism, Clinical and Experimental Research, 38,* 930–937.

48. Yan, J., Aliev, F., Webb, B. T., Kendler, K. S., Williamson, V. S., Edenberg, H. J., et al. (2013). Using genetic information from candidate gene and genome-wide association studies in risk prediction for alcohol dependence. *Addiction Biology, 19,* 708–721.

49. Bakermans-Kranenburg, M. J., & van IJzendoorn, M. H. (2015). The hidden efficacy of interventions: Gene X Environment experiments from a differential susceptibility perspective. *Annual Review of Psychology, 66,* 381–409.

50. Belsky, J., & Pluess, M. (2009). Beyond diathesis stress: Differential susceptibility to environmental influences. *Psychological Bulletin, 135,* 885–908.

51. Hankin, B. L., Nederhof, E., Oppenheimer, C. W., Jenness, J., Young, J. F., Arbela, J.R.Z. et al. (2011). Differential susceptibility in youth: Evidence that 5-HTTLPR × positive parenting is associated with positive affect "for better and worse." *Translational Psychiatry, 1,* e44.

52. Roisman, G. I., Newman, D. A., Fraley, R. C., Haltigan, J. D., Groh, A. M., & Haydon, K. C. (2012). Distinguishing differential susceptibility from diathesis-stress: Recommendations for evaluating interaction effect. *Development and Psychopathology, 24,* 389–409.

53. Boardman, J. D., Daw, J., & Freese, J. (2013). Defining the environment in gene-environment research: Lessons from social epidemiology. *AJPH, 103,* S64–S72.

54. Windle, M. (2010). A multilevel developmental contextual approach to substance use and addiction. *Biosocieties, 5,* 124–136.

55. Zucker, R. A. (2006). Alcohol use and the alcohol use disorders: A developmental-biopsychosocial systems formulation covering the life course. In D. Cicchetti & D. J. Cohen (Eds.), Developmental psychopathology (Vol. 3). *Risk, Disorder and Adaptation* (2nd ed., pp. 620–656). New York: Wiley.

56. Caspi, A., Hariri, A. R., Holmes, A., Uher, R., & Moffitt, T. E. (2010). Genetic sensitivity to the environment: The case of the serotonin transporter gene and its

implications for studying complex diseases and traits. *American Journal of Psychiatry, 167,* 509–527.

57. Binder, E. B., Bradley, R. G., Liu, W., Epstein, M. P., Deveau, T. C., Mercer, K. B., et al. (2008). Association of FKBP5 polymorphisms and childhood abuse with risk of posttraumatic stress disorder symptoms in adults. *Journal of the American Medical Association, 299,* 1291–1305.

58. Dick, D. M., Latendresse, S. J., Lansford, J. E., Budde, J. P., Goate, A., Dodge, K.A., et al. (2009). Role of *GABRA2* in trajectories of externalizing behavior across development and evidence of moderation by parental monitoring. *Archives of General Psychiatry, 66,* 649–657.

59. Latendresse, S. J., Bates, J. E., Goodnight, J. A., Lansford, J. E., Budde, J. P., Goate, A., et al. (2011). Differential susceptibility to adolescent externalizing trajectories: Examining the interplay between *CHRM2* and peer group antisocial behavior. *Child Development, 82,* 1797–1814.

60. Mrug, S., & Windle, M. (2014). DRD4 and susceptibility to peer influence on alcohol use from adolescence to adulthood. *Drug and Alcohol Dependence, 145,* 168–173.

61. Hutter, C. M., Mechanic, L. E., Chatterjee, N., Kraft, P., & Gillanders, E. M. (2013). Gene-environment interactions in cancer epidemiology: A National Cancer Institute Think Tank report. *Genetic Epidemiology, 37,* 643–657.

62. Brody, G. H., Chen, Y. F., Beach, S. R. H., Kogan, S. M., Yu, T., DiClemente, R.J., et al. (2014). Differential sensitivity to prevention programming: A dopaminergic polymorphism-enhanced prevention effect on protective parenting and adolescent substance use. *Health Psychology, 33,* 182–191.

63. Beach, S. R. H., Brody, G. H., Lei, M. K., & Philibert, R. A. (2010). Differential susceptibility to parenting among African American youths: Testing the DRD4 hypothesis. *Journal of Family Psychology, 24,* 513–521.

64. Dick, D. M. (2011). Gene-environment interaction in psychological traits and disorders. *Annual Review of Clinical Psychology, 7,* 383–409.

65. Marjoram, P., Zubair, A., & Nuzhdin, S. V. (2014). Post-GWAS: where next? More samples, more SNPs or more biology? *Heredity, 112,* 79–88.

66. Duncan, L. E., & Keller, M. C. (2011). A critical review of the first 10 years of candidate gene-by-environment interaction research in psychiatry. *American Journal of Psychiatry, 168,* 1041–1049.

67. Duncan, L. E., Pollastri, A. R., & Smoller, J. W. (2014). Mind the gap: Why many geneticists and psychological scientists have discrepant views about gene-environment interaction (GXE) research. *American Psychologist, 69,* 249–268.

68. Moffitt, T. E., Caspi, A., & Rutter, M. (2005). Strategy for the investigating interactions between measured genes and measured environments. *Archives of General Psychiatry*, *62*, 473–481.

69. Sher, K. J., Dick, D., Crabbe, J. C., Hutchison, K. E., O'Malley, S. S., & Heath, A. C. (2010). Consilient research approaches in studying gene X environment interactions in alcohol research. *Addiction Biology*, *15*, 200–216.

70. Bosker, F. J., Hartman, C. A., Nolte, I. M., Prins, B. P., Terpstra, P., Posthuma, D., et al. (2011). Poor replication of candidate genes for major depressive disorder using genome-wide association data. *Molecular Psychiatry*, *16*, 516–532.

71. Siontis, K. C., Patsopoulos, N. A., & Ioannidis, J. P. (2010). Replication of past candidate loci for common diseases and phenotypes in 100 genome-wide association studies. *European Journal of Human Genetics*, *18*, 832–837.

72. Guintivano, J., Brown, T., Nexcomer, A., Cox, O., Maher, B. S., Eaton, W.W., et al. (2014). Identification and replication of a combined epigenetic and genetic biomarker predicting suicide and suicidal behaviors. *American Journal of Psychiatry*, *171*, 1287–1296.

73. Kendler, K. S., Myers, J., & Prescott, C. A. (2007). Specificity of genetic and environmental risk factors for symptoms of cannabis, cocaine, alcohol, caffeine, and nicotine dependence. *Archives of General Psychiatry*, *64*, 1313–1320.

74. Iacono, W. G., Malone, S. M., & McGue, M. (2008). Behavioral disinhibition and the development of early-onset addiction: Common and specific influences. *Annual Review of Clinical Psychology*, *4*, 325–348.

75. Windle, M. (in press). Behavioral undercontrol: A multifaceted concept and its relationship to alcohol and substance use and abuse. In R. Zucker & S. Brown (Eds.), *Oxford Handbook of Adolescent Substance Abuse*. Oxford University Press.

76. Arcos-Burgos, M., Velez, J. I., Solomon, B. D., & Muenke, M. (2012). A common genetic network underlies substance use disorders and disruptive or externalizing disorders. *Human Genetics*, *131*, 917–929.

77. Sher, K. J., Grekin, E. R., & Williams, N. A. (2005). The development of alcohol use disorders. *Annual Review of Clinical Psychology*, *1*, 493–523.

78. Sanchez, B. N., Kang, S., & Mukherjee, B. (2012). A latent variable approach to study gene-environment interactions in the presence of multiple correlated exposures. *Biometrics*, *68*, 466–476.

79. Sanchez, B. N., Budtz-Jorgensen, E., Ryan, L. R., & Hu, H. S. (2005). Structural equation models: A review with applications to environmental epidemiology. *Journal of the American Statistical Association*, *100*, 1443–1455.

11 A Statistical Framework for Mediation in Environmental Epigenetics

Duncan C. Thomas

"Epigenetics" refers to factors affecting the expression of genes without changes in the underlying DNA sequence. The best studied is DNA methylation, but other changes include various posttranslational histone modifications, regulation by noncoding microRNAs, and so forth. These mechanisms play a central role in X-chromosome inactivation and genomic imprinting. Our focus is initially on epigenetic changes that can be transmitted mitotically in somatic cells (later we will consider meiotic transmission), potentially providing a mechanism by which environmental exposures can affect gene expression and hence disease risk or other traits. Indeed, it has been demonstrated that MZ twins (who are genetically identical) have similar DNA methylation and histone acetylation patterns—both globally and at specific loci—early in life, but acquire remarkably large differences as they age [1], possibly in response to different exposure histories but also possibly just due to random drift. However, heritability of DNA methylation may depend on the locus in question. Recent work evaluating heritability of DNA methylation in *AXL*, a receptor tyrosine kinase relevant in cancer and immune function, found a low level of heritability in twins as young as nine years of age [2]. Ever since Barker [3] proposed a fetal basis of adult disease (the "Barker hypothesis"), a broad range of epidemiologic research has implicated various environmental exposures during prenatal and early postnatal development as influencing the risk of adult cardiovascular disease, obesity, type 2 diabetes, and other chronic diseases. More recently, de Boo and Harding [4] and Waterland and Jirtle [5] have discussed the evidence that epigenetics may provide the mechanistic basis for these associations.

It is known that oxidative stress is capable of producing epigenetic changes [6]. For example, long-term exposure to PM_{10} has been associated with reduced methylation of Alu and LINE-1 elements as well as iNOS in humans [7] and PM_{10}, $PM_{2.5}$, and ozone exposures are associated with DNA methylation of iNOS [2]. In the Southern California Children's Health Study, a common promoter haplotype in *NOS2A* (which encodes iNOS) has been found to interact with short-term term $PM_{2.5}$ exposure to influence iNOS methylation and there is a three-way interaction among *NOS2A* haplotypes, iNOS methylation, and $PM_{2.5}$ exposure on exhaled nitric oxide (FeNO, a biomarker of airway inflammation) [8]. In vitro studies have demonstrated associations of DNA methylation with various metals [9, 10]. Exposure to maternal smoking during pregnancy has also been associated with reduced methylation of Alu, of LINE-1 among children with *GSTM1* null genotype, and with increased methylation in the promoters of *AXL* and *PTPRO* [11]. In Greenland Inuit, who have some of the highest levels of persistent organic pollutants, an inverse relationship was found with Alu methylation [12]. Sperm from mice exposed to steel plant air were found to be hypermethylated, a change that persisted long after cessation of exposure [13].

Recently, Huang et al. [14, 15] have proposed a general framework for causal mediation analysis to investigate the effects of SNPs on a disease risk mediated through gene expression, or more generally through methylation and then through expression. Here, we extend this general approach to look at mediation of the joint effects of exposure and genotype and to multiple genes.

Recently it has been recognized that it is also possible for epigenetic changes to be transmitted across generations, potentially affecting disease risks in future generations [16]. For example, transmission of abnormal DNA methylation has been described in two established DNA mismatch repair genes, *MLH1* and *MSH2*, across multiple generations in a few colorectal cancer families in the absence of mutations in these genes [17–19]. The exact mechanism for how such changes can escape the erasure and reestablishment of DNA methylation patterns that occurs during gametogenesis and early embryogenesis remains elusive, but is the focus of intense research. Holliday [20] appears to have been the first to propose the term "epimutation" to describe this phenomenon. It appears that a small subset of genes can escape this reprogramming; the term "metastable epialleles" refers to alleles that are variably expressed through epigenetic changes established early in development [21]. For reviews of the animal and human literature on environmental epigenetics, see [6, 9, 22–29].

The best-documented example of transgenerational determination of a trait through DNA methylation is the determination of coat color and other traits like obesity, diabetes, and tumors in agouti mice. Agouti dams, who carry the metastable A^{vy} epiallele, are more likely to produce agouti offspring while yellow mice (which do not carry that allele) are more likely to produce yellow offspring, with a continuous gradient across the degree of transmission of the A^{vy} epiallele [30, 31]. Furthermore, supplementation of the dam's diet with methyl donors can change the epigenetic state of this allele for several subsequent generations [32–34]. Other experimental evidence of transgenerational epigenetic effects includes exposure of pregnant rats to several endocrine-disrupting agents (vinclozolin, bisphenol A), producing changes in male infertility, body weight, and cancers in up to four subsequent generations [35, 36]. In some cases, these effects can be reversed by dietary supplementation [32, 36].

Whether such phenomena exist in humans remains highly controversial [24]. Several examples of apparent transmission of an environmental insult across generations exist, but none have been convincingly ascribed to epigenetic mechanisms. For example, Pembrey et al. [37] described an association between paternal grandfather's exposure to a famine in northern Sweden and mortality in their grandsons, whereas paternal grandmother's food supply was associated instead with mortality in their granddaughters. They also described effects of preadolescent paternal smoking on BMI in sons but not daughters, while Lumey et al. [38] reported effects on birth weight up to two generations following the Dutch famine in World War II. The Children's Health Study has demonstrated an effect of grandmaternal smoking while pregnant on the asthma risk of their grandchildren, even after controlling for maternal smoking and the child's environmental tobacco smoke exposure [39]. Although any of these phenomena could be due to epigenetic mechanisms, mutation to the DNA itself would be a plausible alternative, as has been shown for mainstream tobacco smoke [40]. Indeed, convincing evidence of transgenerational effects via an epigenetic mechanism would require at least four generations, since the germline of both the offspring and grandchildren are directly exposed to potential mutagenesis during pregnancy [23, 41].

Although there are a few examples of effects of exposure on first- or second-generation offspring, it is presently unknown whether such effects are mediated by epigenetic mechanisms. Bollati and Baccarelli [24] conclude

that "to the best of our knowledge, a formal concept of epigene-environ-
ment interaction has not yet been developed." Perhaps the greatest significance of this line of research was noted by Dolinoy and Jirtle [42]: "unlike
genetic mutations, epigenetic profiles are potentially reversible. Therefore,
epigenetic approaches for prevention and treatment, such as nutritional
supplementation and/or pharmaceutical therapies may be developed to
counteract negative epigenomic profiles." This paper represents an attempt
to provide a statistical framework to address these questions.

Epigenetic Mediation within an Individual

Statistical Methods

Let G denote an individual's constitutional genotype and $G^*(t) = (G, M(t))$
denote the germline genotype G annotated with "epigenetic marks" M
at a particular point in time t, and let $Y(t)$ denote the subject's phenotype
at time t. Epigenetic marks can be highly multidimensional, e.g., beta
values for percent methylation at specific CpGs in promoter regions of
various candidate genes in particular tissues, but for the purpose of exposition, we will treat them as a single continuous variable ranging between
0 and 1. Likewise, G may consist of multiple loci, but for now it will
be treated as a single diallelic locus. We are interested in studying the way
$M(t)$ depends upon an individual's exposure history $E(t)$ (which could
of course include in utero exposures) and possibly genotype and other
factors, and how $M(t)$ in turn influences disease risk, together with $E(t)$, G,
and other risk factors. In particular, in the next section, our interest is in
assessing the extent to which $M(t)$ mediates the relationship between $E(t)$
and $Y(t)$. In a later section, we consider how M can be transmitted across
generations and thereby potentially mediate transgenerational exposure
effects.

Our approach to mediation within an individual entails jointly modeling the epigenetic process and phenotype (figure 11.1). Here, $Y(t)$ could be
a simple cross-sectional observation at a single time point t, longitudinal
measurements of a continuous phenotype (e.g., FeNO), or censored time-
to-event data (e.g., asthma incidence) $Y = (\delta, t)$ where δ is disease status
(1 = case, 0 = unaffected) and t is event or censoring time. Following an
approach we introduced earlier [43], which was subsequently extended by
others [44–48], we propose two models as follows:

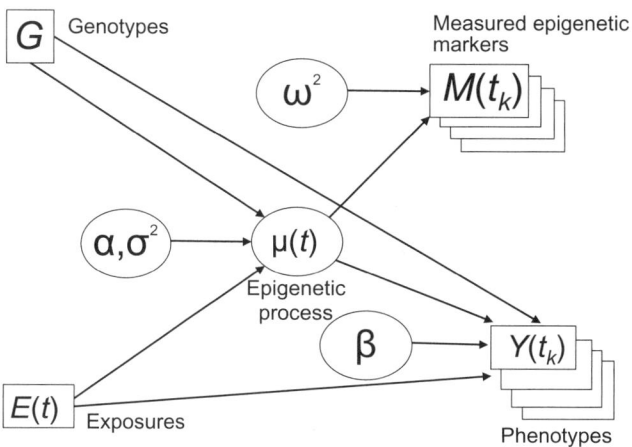

Figure 11.1
Schematic representation of the model for mediation of exposure effects within the individual (boxes represent observed data, circles latent variables or model parameters).

Epigenetic process: $\varepsilon[M(t)] = s(t) + \alpha_1 E(t) + \alpha_2 G + \alpha_3 G \times E(t)$

$\mathrm{cov}[M(t), M(u)] = \sigma^2 \exp(-\rho|t-u|)$

Phenotype process, depending on the nature of the trait, e.g.,

binary disease: $\mathrm{logit}\,\Pr(Y=1) = \beta_1 E + \beta_2 G + \beta_3 G \times E + \beta_4 M$

time to event: $\lambda(t) = \lambda_0(t)\exp[\beta_1 E(t) + \beta_2 G + \beta_3 G \times E(t) + \beta_4 M(t)]$

longitudinal data: $\varepsilon[Y(t_k)] = \beta_1 E(t_k) + \beta_2 G + \beta_3 E(t_k)G + \beta_4 M(t_k)$

These are intended simply as generic examples of possible model forms. For example, the epigenetic process is described as a linear model in exposure, genotype, and their product, with a smooth in time and an AR-1 covariance process, but it may be appropriate to use a general linear model (perhaps with a logit link function if $M(t)$ is restricted to values in $(0,1)$, add various temporal interaction terms, or a more general covariance structure. In both models, it may be appropriate to define the exposure variable $E(t)$ as some measure of cumulative or average exposure up to time t or between intervals of observation. In practice, $M(t)$ is not continuously observed, so we introduce a continuous latent process $\mu(t)$, treating the observed data as $M(t_k) \sim N(\mu(t_k), \omega^2)$ and replace $M(t)$ by $\mu(t)$ in the preceding models. For simplicity, we have also not included possible interactions of M (or μ) with E, G, or both, although these could easily be added.

In some situations where the various parts of the model are all conjugate, it may be possible to fit it directly by maximum likelihood, but in many cases is may be more convenient to use Markov chain Monte Carlo (MCMC) methods. This entails iteratively sampling the parameters (α, σ^2) of the underlying epigenetic process $\mu(t)$ given the current values of β and ω^2 and then updating the latter treating the $\mu(t)$ as known. Details of the model fitting process are provided in the appendix.

Causal Mediation Analysis Methods

In linear models, mediation can be tested by comparing the direct estimate of $\tilde{\beta}_1$ from the model that omits M (or μ) with the joint estimate $\hat{\beta}_1$ when M (or μ) is included in the model. The mediated effect size is then given by either the product $\hat{\alpha}_1 \hat{\beta}_4$ or the difference $\hat{\beta}_1 - \tilde{\beta}_1$ (and similarly for the G main effect or the G × E interaction effect), the two estimators being equivalent for linear models [49–51]. For nonlinear models, however, this does not apply, so the methods of "causal mediation analysis" are required [52–54]. Briefly, for a dichotomous, time-constant exposure E, these would entail computing the "average direct causal effect (ADCE)" as

$$ADCE = \varepsilon_{M|E=0}\left[\varepsilon(Y \mid E = 1, M) - \varepsilon(Y \mid E = 0, M)\right]$$
$$= \int \left[\varepsilon(Y \mid E = 1, M) - \varepsilon(Y \mid E = 0, M)\right] dp\,(M \mid E = 0)$$

and the average mediated causal effect (AMCE) as

$$AMCE = \varepsilon(Y \mid E = 1), \varepsilon(M \mid E = 1) - \varepsilon(Y \mid E = 1), \varepsilon(M \mid E = 0)$$
$$= \int E(Y \mid E = 1, M)\left[\varepsilon(M \mid E = 1) - \varepsilon(M \mid E = 0, M)\right] dp\,(M \mid E = 1)$$

The situation of continuous, time-varying exposures and mediators is considerably more complex and generally would require the use of G-computation methods [54]. Implementation of this approach in the context of longitudinal methylation measurements and time-to-event data for disease is described in the simulation section that follows.

Simulation Studies

We simulated data for 1,000 subjects under a simplified version of models (11.1) and (11.2) with a time-constant exposure, no genetic covariates, a linear growth curve model for the underlying methylation process with covariances induced by random subject effects on slopes and intercepts,

Unobserved methylation process: $\mu_i(t) = (\alpha_0 + \alpha_1 E_i + a_{i1}) + (\alpha_2 + \alpha_3 E_i + a_{i2})t$

$$(11.1)$$

Longitudinal measurements: $M_{ij} \sim N(\mu_i(t_j), \sigma^2)t_j = (0,1,2,4,8,16,32,64)$

$$(11.2)$$

Time to disease onset: $\lambda_i(t) = \lambda_0(t)\exp(\beta_1\mu_i(t) + \beta_2 E_i)$ (11.3)

where $\lambda_0(t) = \exp(\lambda_1 + \lambda_2 t)$. The environmental variable E was generated as lognormally distributed with a log mean of 1 and a log SD of 0.5. Censoring times were generated with rates $v(t) = \exp(-5 + 1.5t)$ and times to event or censoring were truncated at age 99. These values, and the parameters of interest shown in table 11.1, were adjusted to yield an average of approximately 500 cases and 500 controls. Results were based on 10,000 MCMC samples, after discarding a burn-in of 5,000 iterations. The entire process was replicated 100 times.

Trace plots (not shown) indicated adequate mixing. Most of the parameters were well estimated and the mean of the posterior variance estimates agreed well with the empirical variances of the replicate-specific estimates (table 11.1). The estimates of the regression coefficients for the methylation process and its measurement error variance were more precisely estimated than the parameters of the disease process (baseline hazards and relative risks).

Table 11.1
Simulation results for the simplified model for time-varying methylation and time-to-event data for a disease outcome

Parameter	Interpretation	Simulated value	Posterior mean	Posterior SD	Empirical SD across replicates
α_0	μ intercept	3	2.9965	0.0289	0.0271
α_1	$\mu \mid E$	1	0.9991	0.0364	0.0362
α_2	$\Delta\mu$ intercept	1	0.9993	0.0136	0.0133
α_3	$\Delta\mu \mid E$	0.5	0.4990	0.0178	0.0173
β_0	RR $\mid \mu$	1	1.019	0.099	0.091
β_1	RR $\mid E$	1	0.997	0.116	0.114
λ_0	ln λ intercept	−10	−10.08	0.36	0.33
λ_1	ln $\lambda \mid t$	3	3.01	0.15	0.14
σ^2	var(M $\mid \mu$)	1	1.0011	0.0083	0.0083

This simple simulation did not include any genetic determinants, but they could readily be added as additional covariates like the E_i variables in the methylation and/or disease models (along with interaction effects and any other fixed covariates) without introducing any fundamentally new concepts. More complex issues would arise if the exposure is time-dependent or if the disease risk depends on some function of the past history of exposure or methylation levels. Of course, any of these models could also be nonlinear, e.g., a logit-normal model for methylation or a more complex covariance function.

We used the MCMC iteration outputs to estimate the mediation quantities, averaging over the posterior distribution of model parameters and random effects $\theta = (\alpha, \beta, \lambda, \{a_i\})$ and across subjects. At each iteration, we generated random counterfactual event times under four scenarios, increasing and decreasing exposure by $\Delta E/2$ while holding $\mu(t)$ fixed, and generating new $\mu(t)$ trajectories using these modified $E^{\pm}$ values. Let T_{ij}^{em} be a randomly generated event time for subject i using the jth iteration estimates of the hazard rate

$$\lambda_{ij}^{em} = \lambda \left[t \mid E_i^e, \mu \left(t \mid E_i^m; \alpha_j, a_{ij} \right); \beta_j, \lambda_j \right]$$

where e and m index either the observed exposures E_i or the counterfactual ones E_i^+ or E_i^-. Keeping the observed censoring times fixed, we then generated the corresponding failure indicators Y_{ij}^{em} and computed the person- and iteration-specific effect estimates $D_{ij} = Y_{ij}^{+0} - Y_{ij}^{-0}$ for direct effects and $I_{ij} = Y_{ij}^{0+} - Y_{ij}^{0-}$ for indirect effects. (Because of the nonlinearity of the model, the total effect $Y_{ij}^{++} - Y_{ij}^{--}$ is slightly different from the sum of the direct and indirect effects.) The average direct causal effect (ADCE) is then $\hat{D} = \varepsilon_{ij}(D_{ij}) = \varepsilon_j(\bar{D}_j)$ where $\bar{D}_j = \varepsilon_i(D_{ij})$ is the estimate of the ADCE estimate for all subjects at the jth iteration. There are two variance components of interest: between subjects $V_S^D = \varepsilon_j[\text{var}_i(D_{ij})]$ and over the posterior distribution of model parameters $V_\theta^D = \text{var}_j[\bar{D}_j]$, and similarly for the indirect effects. Because the phenotype here is a dichotomous disease status, we dispensed with the between-subjects variance, but computed a similar quantity across replicate datasets. With the same model parameters used to generate simulations in table 11.1, this yielded unit effect estimates (i.e., for $\Delta E = 1$) of $\hat{D} = 0.0594$ with variance components $V_S^D = 0.0067^2$ and $V_\theta^D = 0.0097^2$ for direct effects and yielded $\hat{I} = 0.0954$ with variance components $V_S^I = 0.0098^2$ and $V_\theta^I = 0.0118^2$ for indirect effects.

Application to the ARG-NOS Pathway

Complete data on single nucleotide polymorphisms and promoter methylation in the *ARG2* and *NOS2A* genes, along with exposures to ambient $PM_{2.5}$ and exhaled nitric oxide (FeNO) were available for 114 asthmatic and 706 nonasthmatic subjects from the Children's Health Study [2, 11, 55–58]. We hypothesized that FeNO measurements were determined by expression of these two genes, and that these effects might be modified by asthma status. Gene expression levels were not directly measured for this study but were postulated to depend in turn on genotype and methylation level of the respective genes and that methylation of each gene would depend upon air pollution exposure (figure 11.2). As originally conceived, the full model was given by the following system of equations, letting subscripts $j = 1,2$ denote the two genes:

Methylation: $M_{ij} \sim N\left(\alpha_{0j} + \alpha_{1j}X_i,\ \sigma_{Mj}^2\right)$

Expression: $E_{ij} \sim N\left(\gamma_{1j}G_i + \gamma_{2j}M_{ij},\ \sigma_{Ej}^2\right)$

Inflammation: $L_i \sim N\left(\delta_{0j} + \delta_1 E_{i1} + \delta_2 E_{i2},\ \sigma_I^2\right)$

FeNO biomarker: $B_i \sim N\left(L_i,\ \sigma_B^2\right)$

Incident asthma: $\text{logit}\,\Pr(Y_i = 1) = \beta_0 + \beta_3 L_i + \beta_1 G_{i1} + \beta_2 G_{i2}$

However, only a small number of incident asthmatic cases were included in this dataset, the majority being prevalent cases at the start of follow-up. Since asthma status has a strong effect on FeNO and also appears to modify the effect of genetic and environmental risk factors, we revised the model to treat the biomarker the dependent variable and asthma status as a risk factor. This also had the consequence of making the latent inflammatory variable L nonidentifiable. Hence, we rewrote the last three equations as

FeNO biomarker:

$$B_i \sim N\left(\delta_0 + \delta_{11}E_{i1} + \delta_{21}E_{i2} + \delta_{12}E_{i1}Y_i + \delta_{22}E_{i2}Y_i + \eta_0 X_i + \eta_1 G_{i1} + \eta_2 G_{i2} + \eta_3 Y_i,\ \sigma_I^2\right)$$

where the δ terms indicate the mediated effects and the η terms the direct effects.

Fitting the full model requires certain additional constraints on the parameters to achieve identifiability, since, for example, doubling the dependence of the unobserved expression levels on methylation and genotype while halving the dependence of FeNO on expression would yield

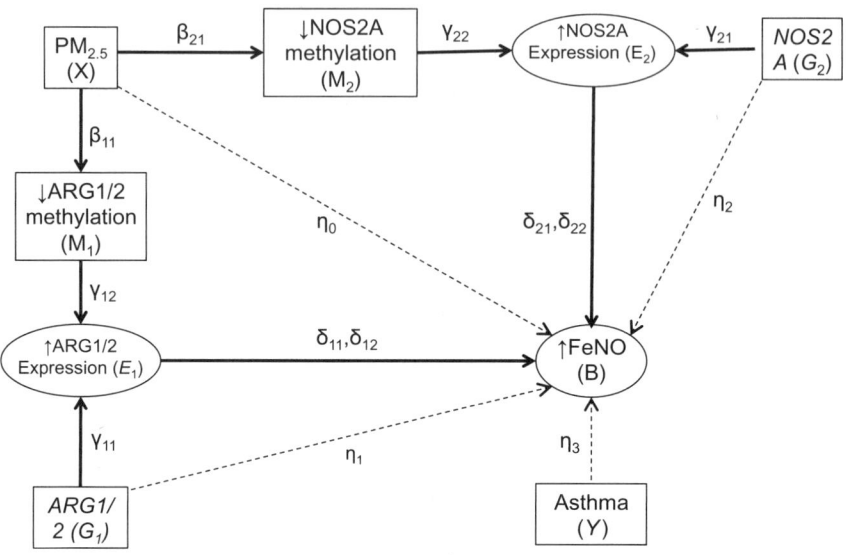

Figure 11.2
Latent variables formulation of the ARG-NOS pathway in the Children's Health
Study data.

identical results. Absent any latent variables, however, the model would
be identifiable. Hence, rather than applying arbitrary constraints, we relied
on external data about the expression variables that are not observed in
the CHS, specifically from the ABRIDGE consortium [59] on expression in
relation to genotype and methylation. To mimic this situation, we postu-
lated priors on $\gamma_{kj} \sim N(g_{kj}, s_{kj}^2/n)$ and $\sigma_{Ej}^2 \sim IG(n\, s_{Ej}^2, n)$, where g_{kj} and s_{Ej}^2 are
constants fixed at values similar to the estimates of γ_{kj} and σ_{Ej}^2 from the real
data and n is the effective sample size for the ABRIDGE consortium data;
s_{kj}^2 and s_{Ej}^2 were arbitrarily fixed at the estimated variances for the corresponding parameters. Before attempting to fit such a complex model to the
real CHS data, we simulated data with the same structure as the real data,
but with stronger effects. For this purpose, we used the actual exposure,
genotype, and asthma data and generated new methylation, expression,
and FeNO data, and then analyzed these data with the same methods. All
the parameter estimates were close to their simulated values (table 11.2),
demonstrating the validity of the fitting procedure.

Table 11.2
Estimates of the parameters of the simulated ARG-NOS model

Parameter	Interpretation	Gene 1		Gene 2	
		True	Est. (SE)	True	Est. (SE)
β_{g1}	$M \mid X$	−1	−1.003 (.034)	−1	−0.982 (.035)
γ_{g1}	$E \mid M$	−1	−1.005 (.048)	−1	−1.004 (.045)
γ_{g2}	$E \mid G$	1	0.997 (.047)	1	0.993 (.048)
δ_{g0}	$B \mid E, Y = 0$	0.5	0.488 (.044)	0.5	0.489 (.042)
δ_{g1}	$B \mid E, Y = 1 - B \mid E, Y = 0$	1	0.98 (.11)	1	1.02 (.12)
Average causal mediation effects					
$\beta_g \gamma_{g1}\ (\delta_{g0} - \delta_{g1} Y^*)$	$B \mid E(M(X)), Y = 0$	0.361	0.354 (.041)	0.361	0.341 (.039)
$\gamma_{g2}\ (\delta_{g0} - \delta_{g1} Y^*)$	$B \mid E(G), Y = 0$	0.361	0.351 (.044)	0.361	0.344 (.044)
$\beta_g \gamma_{g1}\ [\delta_{g0} + \delta_{g1}(1 - Y^*)]$	$B \mid E(M(X)), Y = 1$	1.361	1.338 (.123)	1.361	1.346 (.122)
$\gamma_{g2}\ [\delta_{g0} - \delta_{g1}(1 - Y^*)]$	$B \mid E(G), Y = 1$	1.361	1.326 (.119)	1.361	1.357 (.133)
Direct effects					
η_3	$B \mid Y$	1	1.06 (.21)		
η_0	$B \mid X$	0.5	0.539 (.065)		
η_g	$B \mid G$	0.5	0.503 (.061)	0.5	0.534 (.061)

Models for High-Dimensional Methylation Data

To address highly multidimensional epigenetic data (e.g., multiple CpG sites or islands within genes, multiple genes), we extend previous profile clustering approaches [60–64] to incorporate genetic and environmental determinants and to test the mediation of exposure-response relationships through this probabilistic classification of methylation. For simplicity, we ignore for now the temporal element and assume that exposure, epigenetics, and outcomes are measured at a single point in time. The basic model is illustrated in figure 11.3, following a framework we previously developed for relating haplotypes to disease by clustering them on their ancestral similarities [65–67]. We let $c = 1 \ldots C$ index the epigenetic profiles, and M now represents the vector of all the epigenetic data. The number of clusters C is not fixed, but assumed to have some prior distribution, such as a truncated Poisson (with zero probability for no clusters) with parameter λ to be estimated (for simplicity this part of the model is not shown in the figure). Individuals are assigned to clusters based on their epigenetic data and their determinants G

and E by sampling from their full conditional probabilities $\Pr(c|M,G,E,C) \propto$ $\Pr(c|\pi(G,E),C) \times \Pr(M|c;\mu,\Sigma)$ where $\pi(G,E) = [\pi_c(G,E)]_{c=1,\ldots,C}$ and

$$\pi_c(G,E) = \frac{\exp(\alpha_{0c} + \alpha_{1c}G + \alpha_{2c}E + \alpha_{3c}G \times E)}{\sum_{c=1}^{C} \exp(\alpha_{0c} + \alpha_{1c}G + \alpha_{2c}E + \alpha_{3c}G \times E)} \tag{11.4}$$

with $\alpha_{01} = \alpha_{11} = \alpha_{21} = \alpha_{31} = 0$ for identifiability. $\Pr(M|c,\mu,\Sigma)$ is a model for the distribution of the methylation data within each cluster, for example multivariate normal with mean vector μ_c and covariance matrix Σ_c, after suitable transformation of the observed data. $\Pr(Y|C=c,G,E)$ could be any of those described before with $\beta_4 M$ replaced by subscript c on α_0, depending upon the nature of the phenotype, for example,

$$\text{logit } \Pr(Y=1) = \beta_{0c} + \beta_1 E + \beta_2 G + \beta_3 GxE. \tag{11.5}$$

Again, mediation is tested by comparing with the null model $H_0:\beta_{0c} \equiv \beta_0$, and interaction effects of c with E or G can be added. The model can be fitted using MCMC methods, either reversible jump to allow for the changing dimension of C or more simply by fixing a maximum number of clusters and allowing some to be empty.

Simulation Studies

We simulated 1,000 subjects, generating G and E data by random sampling from Mendelian(0.5) and $LN(0,1)$ distributions respectively, assigning $C =$

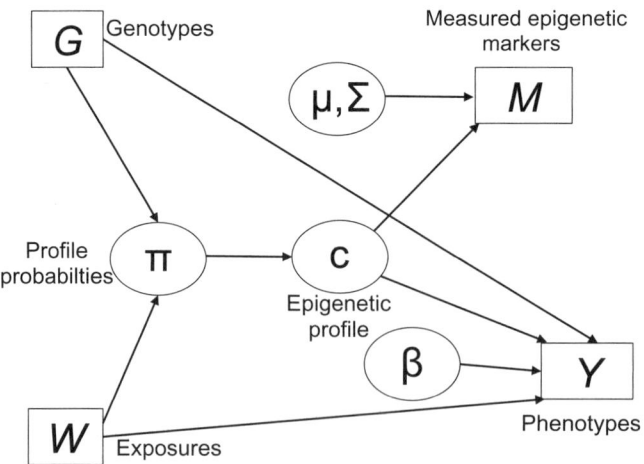

Figure 11.3
Latent class model for multidimensional epigenetic marks.

Table 11.3

Illustrative simulation results for the profile regression model for high dimensional epigenetic markers as mediators of gene-environment effects on disease

Parameter	Interpretation	Cluster-specific posterior mean (SD, simulated value)		
		Cluster 1	Cluster 2	Cluster 3
α_c	polytomous logistic regression coefficients for $\Pr(C \mid G,E)$	0.00* (0.00, 0.0)	−0.92 (0.50, −2.0)	−6.04 (0.50, −6.0)
α_G		0.00* (0.00, 0.0)	−0.24 (0.26, 1.0)	1.19 (0.35, 2.0)
α_E		0.00* (0.00, 0.0)	0.68 (0.21, 1.0)	2.09 (0.34, 2.0)
$\alpha_{G \times E}$		0.00* (0.00, 0.0)	1.44 (0.31, 1.0)	2.20 (0.33, 2.0)
β_c	logistic regression coefficients for $\Pr(Y{=}1 \mid C,G,E)$	−2.74 (0.38, −2.50)	−1.68 (0.24, −1.5)	−0.67 (0.31, −0.5)
β_G			0.82 (0.15, 0.5)	
β_E			0.65 (0.09, 0.5)	
$\beta_{G \times E}$			−0.21 (0.11, 0.5)	
		Mean (simulated value)		
μ_1	Cluster-specific methylation means	−0.05 (0.00)	0.42 (0.40)	0.84 (0.90)
μ_2		0.23 (0.10)	0.21 (0.30)	0.81 (0.80)
μ_3		0.24 (0.20)	0.28 (0.20)	0.75 (0.70)
μ_4		0.27 (0.30)	0.22 (0.10)	0.63 (0.60)
μ_5		0.38 (0.40)	0.04 (0.00)	0.49 (0.50)
μ_6		0.66 (0.50)	−0.01 (0.00)	0.39 (0.40)
μ_7		0.57 (0.60)	0.03 (0.10)	0.37 (0.30)
μ_8		0.68 (0.70)	0.17 (0.20)	0.24 (0.20)
μ_9		0.72 (0.80)	0.21 (0.30)	0.07 (0.10)
μ_{10}		0.95 (0.90)	0.37 (0.40)	−0.11 (0.00)

* reference category, not estimated

1,2,3 under equation (11.4), and then generating Y under Eq.[5] and **M** ~ $MVN(\mu_C, \Sigma_C)$ for ten methylation markers with the cluster-specific means given in table 11.3 and different covariance structures with unit variances (one independent, one exchangeable with $\rho = 0.25$, one AR-1 with $\rho = 0.5^{|m-m'|}$). The parameters α and β were adjusted to yield roughly equal numbers of subjects assigned to each cluster and expected direct and indirect effects of roughly comparable size for main effects and somewhat smaller for G × E interactions. The parameter estimates are provided in table 11.3 for α, β, and μ, several of which show significant departures from their simulated values (a problem still being investigated). Nevertheless, mediation

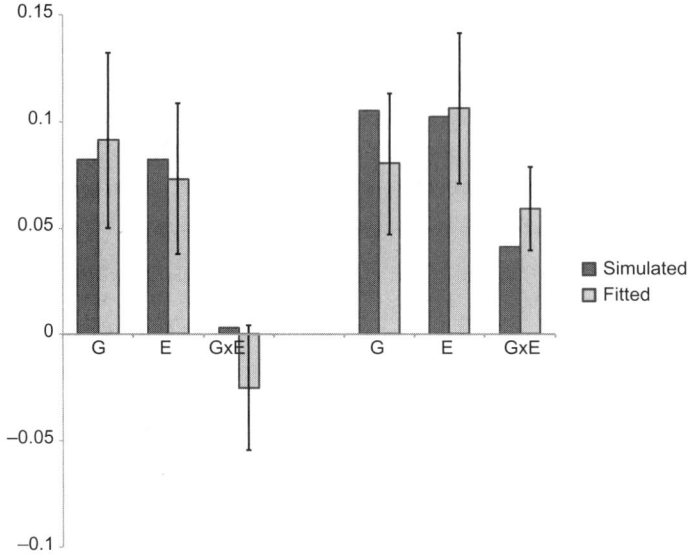

Figure 11.4
Estimates of direct and mediated average causal effects (ADCE and AMCE) in the cluster model.

estimates were generally close to their simulated values (figure 11.4), as were estimated cluster assignments (figure 11.5).

Epigenetic Mediation across Generations

Statistical Methods
We conceptualize the transgenerational effects of exposure mediated through epigenetics as occurring in two stages (figure 11.6). The first is the accumulation of epigenetic marks on the parental germline (denoted by μ) in response to that individual's own exposure history, his or her inherited epigenetic state (μ^T), and possibly genotype at relevant loci. The second is the transmission of these epigenetic marks to the offspring (μ^T) given the parental epigenotype (μ) at the time of conception. For purposes of exposition, we assume it is only the epigenetic marks on the parental transmitted haplotypes that are relevant to determining the offspring's epigenotype and any marks on the nontransmitted haplotypes are lost. So let $G = (H_m, H_f)$ denote the maternally and paternally derived alleles or haplotypes

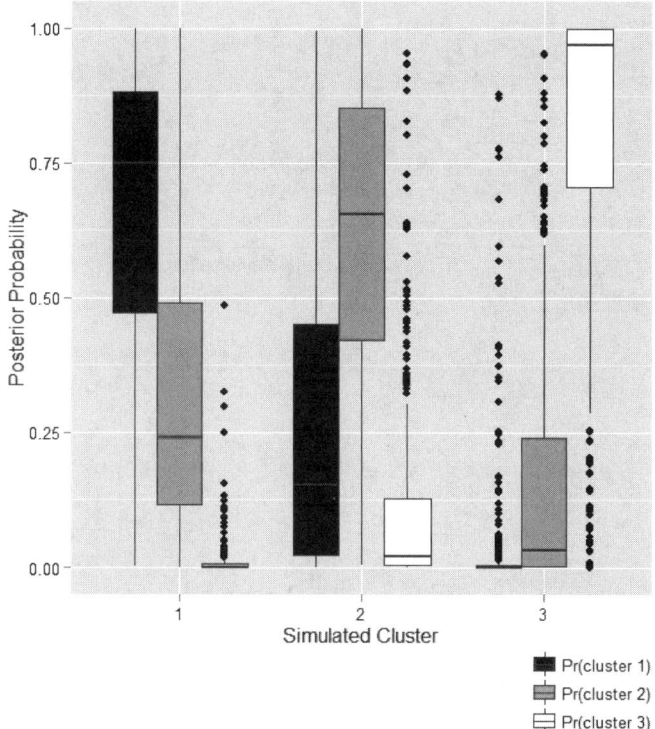

Figure 11.5
Fitted cluster assignment probabilities from the cluster model.

at a candidate locus, and let $G^* = (H_m, \mu_m; H_f, \mu_f)$ include the epigenetic marks on the corresponding haplotypes. We propose a model for transmission of G_p^* to G_o^* from a parent p ($\in \{m,f\}$) to an offspring o as follows. Let $T_p \in \{m,f\}$ denote the grandparental source of the haplotype transmitted from parent p to offspring o, so the offspring's constitutional genotype can be represented as $G_o = (H_{mTm}, H_{fTf})$. Then we propose a transmission model for the "inherited" epigenomic marks $\mu_o^T = (\mu_{om}^T, \mu_{of}^T)$ of the form $\Pr(\mu_{op}^T) = h(\mu_{op}; \tau)$, where $h(.)$ is some function of the epigenetic state M_{pT} of the transmitted parental gamete, with parameters τ; e.g., $E[\text{logit}(\mu_{op}^T)] = \tau_0 + \tau_1 \mu_{pT}$. In turn, the parental epigenetic states are assumed to depend upon their own constitutional genotypes and exposure through a model similar to the epigenetic process model described in the previous section. After conception, these epigenetic marks can by further modified by the subject's own

a)

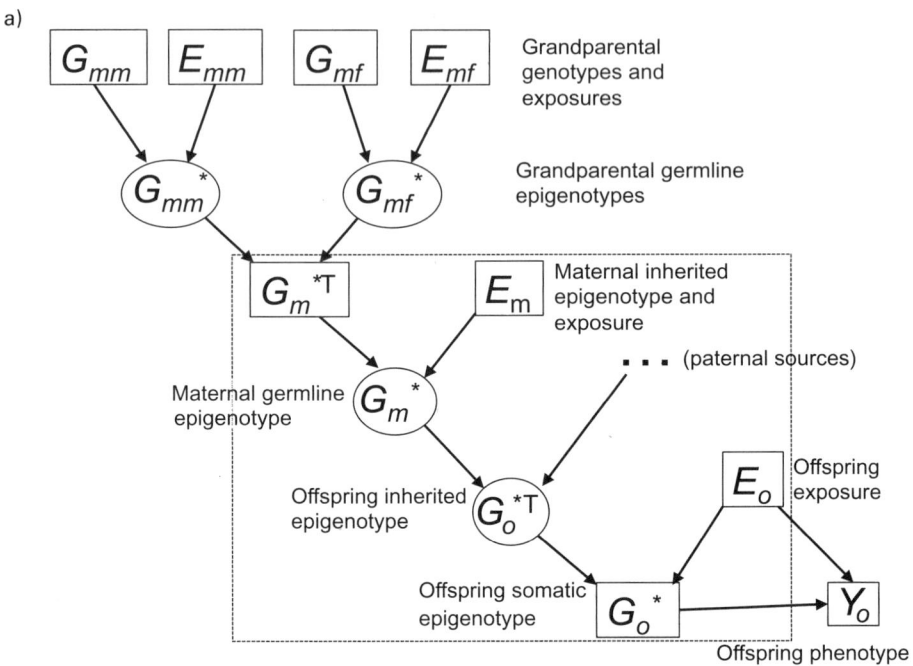

b)

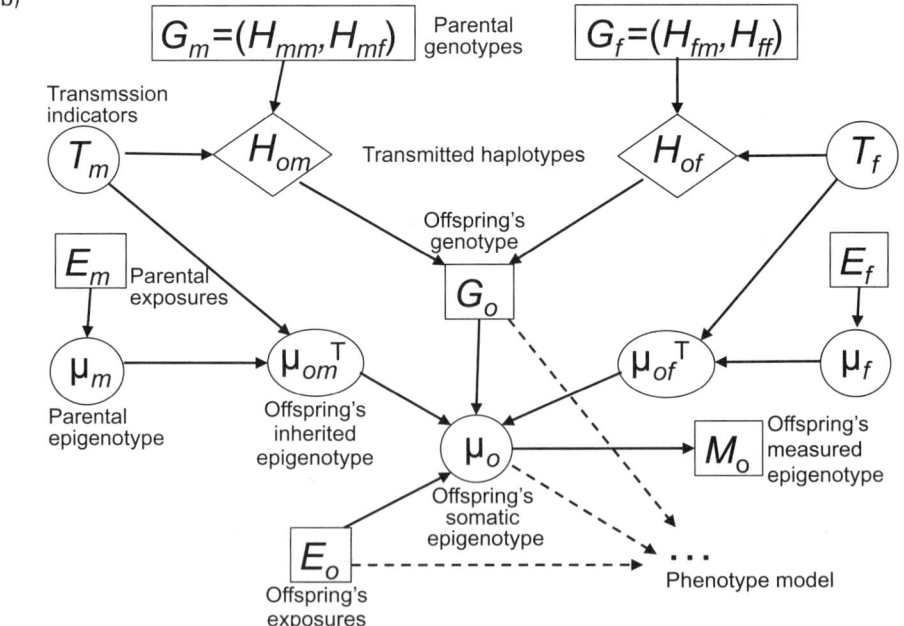

Figure 11.6

(a) Schematic representation of the model for transgenerational effects; (b) detail of the model for transmission across s single generation (the dotted box in the top figure).

exposures (including in utero exposures) to yield a etiologically relevant epigenomic history $\mu(t)$ via a similar model to the previous one, but now with additional terms for the "inherited" methylation state μ^T and possibly its interactions with G or E. Finally, we need a measurement error model for measured epigenetic state $M(t) \sim N(\mu(t), \omega^2)$.

Specifically, we model the latent epigenetic state as

$$\text{logit}(\mu_{ij}) \sim N(\alpha_0 + \alpha_1 E_i + \alpha_2 \mu_{ij}^T, \ \sigma^2),$$ (11.6)

the transmitted epigenetic state as

$$\text{logit}(\mu_{ij}^T) \sim N(\tau_0 + \tau_1 \mu_{pijTij}, \ \varphi^2),$$ (11.7)

and the measured epigenetic state as

$$M_i \sim N((\mu_{i1} + \mu_{i2})/2, \ \omega^2)$$ (11.8)

truncated in $[0,1]$. Finally, the penetrance model is assumed to be

$$\text{logit}[\Pr(Y_i = 1)] = \beta_0 + \beta_1 E_i + \beta_2 G_i + \beta_3 (\mu_{i1} + \mu_{i2})/2.$$ (11.9)

For identifiability, certain constraints are needed. To ensure stationarity across generations, we set $\tau_0 = \alpha_0 + \alpha_1 p_E$ and $\tau_1 = \alpha_2$. The variances σ^2 and φ^2 are not jointly estimable, so they are constrained to be equal. Finally, ω^2 is only estimable jointly with the other variances if there are replicate measures of M_i. This model can also be fitted by MCMC methods. Essentially, each of the unobserved quantities denoted by circles is sampled conditionally on the observed data and the current assignment of the remaining latent variables and model parameters, and then the model parameters are updated conditional on the current assignment of the latent variables. The nine regression coefficients (α, τ, γ) and three variances $(\sigma^2, \varphi^2, \omega^2)$ were estimated by a Metropolis-Hastings MCMC algorithm, along with the unobserved $(T_{ij}, \mu_{ij}, \mu_{ij}^T)$. Details are provided in the appendix.

Simulation Studies

We generated 10 replicate datasets, each comprising 1,000 three-generation families with a single Mendelian gene G, a binary E (with no familial correlation), and a binary phenotype Y. The baseline parameter choices are given in table 11.4, along with the means and standard deviations of the parameter estimates and the mean of their standard error estimates. Except for the parameters τ_1 and φ^2, most of the parameters were well estimated,

Table 11.4
Simulated estimates and standard errors of the parameters of the transgenerational model for ten replicate datasets for a single choice of parameters, 1,000 three-generation pedigrees, as shown in figure 11.1

Parameter	Interpretation	Simulated value	Mean estimate	Mean SE (est.)	SD (est's)
α_1	$E(\mu \mid E)$	0.50	0.46	0.029	0.117
α_2	$E(\mu \mid \mu^T)$	1.00	0.92	0.062	0.199
τ_1	$E(\mu^T \mid \mu)$	1.00	0.49	0.040	0.226
β_1	$E(Y \mid E)$	0.25	0.14	0.102	0.291
β_2	$E(Y \mid G)$	0.50	0.50	0.042	0.207
β_3	$E(Y \mid \mu)$	2.00	2.54	0.409	0.564
σ^2	$\mathrm{var}(\mu \mid E,\mu^T)$	0.20	0.16	0.031	0.121
φ^2	$\mathrm{var}(\mu^T \mid \mu)$	0.20	0.39	0.040	0.153
ω^2	$\mathrm{var}(M \mid \mu)$	0.20	0.19	0.001	0.042

Table 11.5
Minimum detectable relative risks in a study of 1,000 three-generation pedigrees

Comparison	$\Delta\mu$	Minimum detectable RR
Fully methylated vs fully nonmethylated individual	1.00	3.77
$E(\mu \mid G = 1)$ vs $E(\mu \mid G = 0)$	0.24	1.37
$E(\mu \mid G_{par} = 1)$ vs $E(\mu \mid G_{par} = 0)$	0.12	1.17
$E(\mu \mid G_{gp} = 1)$ vs $E(\mu \mid G_{gp} = 0)$	0.07	1.10

although their SE estimates were consistently too small, for reasons that have still not been identified.

The SE of parameter β_3 for the mediating effect of methylation on the phenotype can be used for power calculations. Table 11.5 shows that a RR of 3.77 comparing fully methylated vs. fully nonmethylated states could be detectable with n =1,000 families at a 5 percent two-sided significance level with 90 percent power. A more meaningful comparison would be between the mean methylation status of exposed vs. unexposed individuals ($\Delta\mu = 0.24$), or between the exposure status of their parents ($\Delta\mu = .12$) or grandparents ($\Delta\mu = .07$), for which the corresponding minimum detectable

RRs are also provided in table 11.5. Other parameters of interest are α_2 (SE = 0.016) for the effect of inherited methylation status on the offspring's methylation, α_1 (SE = 0.009) for the effect of exposure on methylation, and τ_1 (SE = 0.067) for the transmission of methylation from parent to offspring.

Discussion

Further development of these simulations might entail primarily adding greater realism. Perhaps most important would be adding in the temporal element to incorporate exposures during the two critical stages of epigenetic reprogramming during gametogenesis and early embryogenesis. Little seems to be known currently about the influence of constitutional genotype or later exposures on epigenetics, but a flexible modeling framework that incorporates these potential effects could be very useful, as would methods for formally testing mediation effects and more complex epigenetic and phenotype models. Our simulation system allows for replicate simulations and the usual MCMC diagnostics for assessing convergence, bias, variance, power, etc. over a range of parameter choices and study designs. Since of course the true mechanism is unknown, further simulations should investigate robustness to model misspecification. The relative informativeness of various study designs is another potentially interesting avenue for further investigation. For example, the results in tables 11.4 and 11.5, based on complete data on all seven members of three-generation pedigrees, could be extended to investigate alternative reduced designs, such as the "pent" design of Weinberg et al., studies with complete environmental data but with genotypes and epigenotypes limited to parents and offspring, and so on. In earlier work, we have explored the optimization of two-phase sampling designs in models involving latent variables like these [68] and pathway models incorporating biomarkers [69].

Potential Cohorts for Investigating Transgenerational Phenomena in Humans

To date, most of the evidence for transgenerational inheritance of exposure effects has come from animal experiments, where it is feasible to follow relatively short-lived animals like rodents for the three or more generations after exposure needed to rule out direct exposure effects to somatic or germ cells while in utero. While several cohorts have been observed to have experienced effects of deleterious exposures in offspring and grand-offspring (e.g., the famine studies mentioned earlier), direct evidence of epigenetic mediation as the mechanism has not been available for lack of DNA samples for methylation analysis, let alone for accurate quantitation of individual's exposure to specific agents. Several multigenerational cohort studies with stored biospecimens are available, however, as reviewed by Cortessis et al. [29], such as the Child Health and Development Study [70], based on a study of pregnancies in the early 1960s, with the third genera-tion currently being enrolled [71–73]. The Framingham family cohort [74, 75] is also now in its third generation. Another opportunity is the Strong Heart Family Study [76, 77], a study of up to four generations in three com-munities of North American Indians impacted by arsenic exposure, which is known to affect methylation and cancer. Finally, the third generation of women exposed in utero to diethylstilbestrol [78, 79] is another promising opportunity. All these studies will require the development of novel statisti-cal methods to fully explore these fascinating data.

Appendix: MCMC Fitting Algorithm

Longitudinal Model
Parameters to be estimated in this model are the coefficients α of the epi-genetic process $[\mu(t) \mid E, \mathbf{a}; \alpha]$ (equation [11.1]), the mean and covariance matrix of the random effects $\mathbf{a}$, the error variance σ^2 of the measurement process $[M(t_k) \mid \mu(t_k)]$ (equation [11.2]), and the baseline risk $\lambda_0(t)$ and log relative risk coefficients β of the disease process (equation [11.3]). In addi-tion, at each cycle of the MCMC fitting process, we update the random effects $\mathbf{a}_i$ for the individual subject's growth curves. Both of these steps are accomplished by sampling from their full conditional distributions using the Metropolis-Hastings algorithm.

Clustering Model
The MCMC iterations proceeded in two alternating steps, sampling individ-uals' cluster assignments using the current model parameters, followed by updating the model parameters conditional on the current cluster assign-ments. The individual assignments were based on

$$[C_i \mid G_i, E_i, Y_i, M_i; \theta] \propto [C_i \mid G_i, E_i,; \alpha] \times [Y_i \mid C_i, G_i, E_i; \beta] \times [M_i \mid C_i; \mu_c, \Sigma_c],$$

where the first two densities are given by equations (11.4) and (11.5), and the third is a multivariate normal density. We simply calculate these probabilities for each possible genotype and choose a cluster for that individual accordingly. Model parameters were updated one at a time using the Metropolis-Hastings algorithm with a normal proposal kernel (with variance calibrated to achieve approximately a 40 percent acceptance rate for each parameter) using the likelihood function

$$L(\alpha,\beta,\mu,\Sigma) = \prod_i p_\beta(Y_i \mid C_i, G_i, E_i) \, p_\alpha(C_i \mid G_i, E_i) p_{\mu,\Sigma}(M_i \mid C_i)$$

Joint updating of model parameters could be done in a similar manner but would require more effort to choose an appropriate proposal interval. The results in table 11.3 are based on fixed number (3) of clusters and 10 methylation markers; possible extension would include using reversible jump MCMC to allow for an unknown number of clusters and semiparametric Bayes models for greater flexibility in modeling the joint distribution of methylation markers.

Mediation estimates were calculated in a similar manner to that described for the longitudinal model, except that cluster assignments were treated as fixed for the *ADCE* and the *AMDE* was calculated by summing over all clusters using the cluster probabilities under the counterfactual alternatives:

$$ADCE = E(Y \mid G'', E'', C) - E(Y \mid G', E', C)$$

$$AMDE = \Sigma_c E(Y \mid G, E, c) \, Pr(C = c \mid G'', E'') - E(Y \mid G, E, c) \, Pr(C = c \mid G', E')$$

averaging over subjects and MCMC iterations, with the indicated expectations and probabilities evaluated based the current model parameter estimates.

Transgenerational Model

The model is fitted by a series of Monte Carlo updates of each latent variable and parameter in turn, sampling from their respective full conditional distributions (figure 11.7).

The transmission indicators are updated using standard Mendelian probabilities:

$$[T_{ij} \mid G_{ij}, G_{pij}] = 1/2 \text{ if } G_{ij} = G_{pijT_{ij}} \text{ and } G_{pij1} = G_{pij2}$$

$$1 \text{ if } G_{ij} = G_{pijT_{ij}} \text{ and } G_{ij1}{}^1 G_{ij2}$$

$$0 \text{ otherwise,}$$

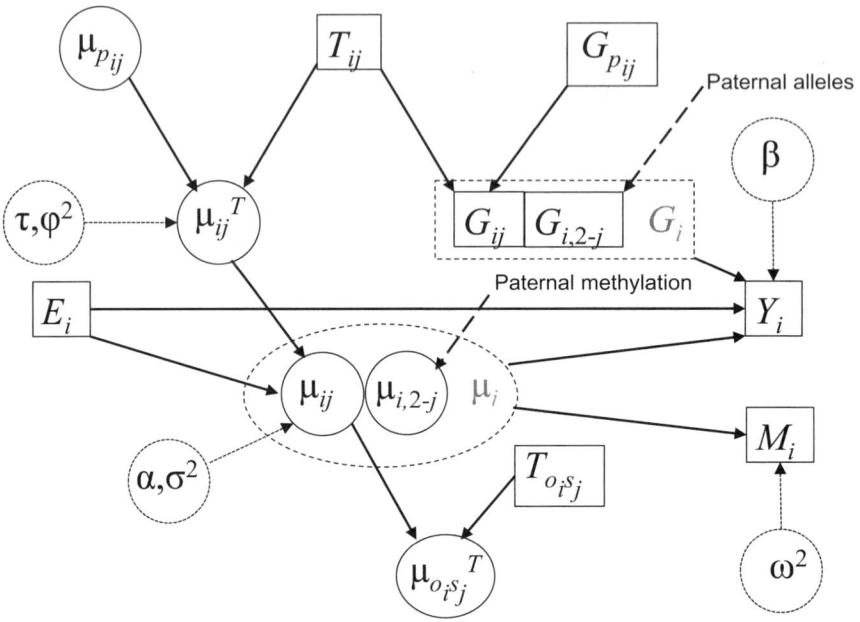

Figure 11.7
DAG showing relationships of all parameters (dotted circles) and latent variables (solid circles) in the transgenerational epigenetic model; dashed boxes and ovals represent joint nodes for genotypes G_i and epigenotypes μ_i on both chromosomes, respectively.

so conditional on the other variables, the full conditional distribution becomes

$$[T_{ij} \mid -] \sim \left[G_{ij} \mid G_{pij}, T_{ij}\right] \left[\mu_{ij}^{T} \mid \mu_{pijTij}\right]_{\tau,\varphi}$$

where the second factor is given by equation (11.7) and "–" indicates conditional on everything else. In a similar manner, the full conditional distributions for the individual and transmitted epigenetic states are given by

$$[\mu_{ij}^{T} \mid -] \sim \left[\mu_{ij}^{T} \mid \mu_{pijTij}\right]_{\tau,\varphi} \left[\mu_{ij} \mid \mu_{ij}^{T}, E_i\right]_{\alpha,\sigma}$$

$$[\mu_{ij} \mid -] \sim \left[\mu_{ij} \mid \mu_{ij}^{T}, E_i\right]_{\alpha,\sigma} \left[\mu_{oisj}^{T} \mid \mu_i, T_{oisi}\right]_{\tau,\varphi} \left[Y_i \mid E_i, G_i, \mu_i\right]_{\beta} \left[M_i \mid \mu_i\right]_{\omega}$$

where the various factors are given by equations (11.6–11.9) (here o_i denote the index of the offspring of subject i, and s_i indicates whether subject i is a mother or father). Each of these updates are performed using a Metropolis-

Hastings move, proposing a small deviation from the current value and then accepting or rejecting the proposal according to the Hastings ratio.

The parameters are updated conditional on the current assignment of the latent variables. The variance parameters are updated by sampling from inverse gamma distributions with scale parameters $\Sigma_{ij} \left[\mu_{ij} - E\left(\mu_{ij} \mid \mu_{ij}^{T}, E_i; \alpha\right) \right]^2$ (for σ^2), $\Sigma_{ij} \left[\mu_{ij}^{T} - E\left(\mu_{ij}^{T} \mid \mu_i, T_{ij}; \alpha\right) \right]^2$ (for φ^2), and $\Sigma_{ij} \left[M_i - E\left(M_i \mid \mu_i\right) \right]^2$ (for ω). The regression parameters are updated using Metropolis-Hastings moves.

Results in table 11.3 are based on 2,000 MCMC iterations for each data replicate, after discarding 1,000 for burn-in. Examination of convergence plots seems to indicate that this number is sufficient to achieve stability.

Acknowledgments

Supported in part by NIH grants ES020794, ES019876, ES07048, CA148107, and CA014089.

References

1. Fraga, M. F., Ballestar, E., Paz, M. F., Ropero, S., Setien, F., Ballestar, M. L., et al. (2005). Epigenetic differences arise during the lifetime of monozygotic twins. *Proceedings of the National Academy of Sciences of the United States of America, 102,* 10604–10609.

2. Breton, C. V., Salam, M. T., & Gilliland, F. D. (2011). Heritability of DNA methylation in AXL receptor tyrosine kinase. *Epigenetics, 6,* 895–898.

3. Barker, D. J., Gluckman, P. D., Godfrey, K. M., Harding, J. E., Owens, J. A., & Robinson, J. S. (1993). Fetal nutrition and cardiovascular disease in adult life. *Lancet, 341,* 938–941.

4. de Boo, H. A., & Harding, J. E. (2006). The developmental origins of adult disease (Barker) hypothesis. *Australian and New Zealand Journal of Obstetrics and Gynaecology, 46,* 4–14.

5. Waterland, R. A., & Jirtle, R. L. (2004). Early nutrition, epigenetic changes at transposons and imprinted genes, and enhanced susceptibility to adult chronic diseases. *Nutrition (Burbank, Los Angeles County, Calif.), 20,* 63–68.

6. Baccarelli, A., & Bollati, V. (2009). Epigenetics and environmental chemicals. *Current Opinion in Pediatrics, 21,* 243–251.

7. Tarantini, L., Bonzini, M., Apostoli, P., Pegoraro, V., Bollati, V., Marinelli, B., et al. (2009). Effects of particulate matter on genomic DNA methylation content and iNOS promoter methylation. *Environmental Health Perspectives, 117,* 217–222.

8. Salam, M. T., Byun, H.-M., Breton, C. V., Wang, X., Eckel, S. P., & Gilliland, F. D. (2011). *Genetic and Epigenetic Variations in Inducible Nitric Oxide Synthase Promoter Region, Particulate Pollution and Exhaled Nitric Oxide in Children.* American Thoracic Society.

9. Dolinoy, D. C., Weidman, J. R., & Jirtle, R. L. (2007). Epigenetic gene regulation: Linking early developmental environment to adult disease. *Reproductive Toxicology (Elmsford, N.Y.), 23,* 297–307.

10. Wright, R. O., & Baccarelli, A. (2007). Metals and neurotoxicology. *Journal of Nutrition, 137,* 2809–2813.

11. Breton, C. V., Byun, H. M., Wenten, M., Pan, F., Yang, A., & Gilliland, F. D. (2009). Prenatal tobacco smoke exposure affects global and gene-specific DNA methylation. *American Journal of Respiratory and Critical Care Medicine, 180,* 462–467.

12. Rusiecki, J. A., Baccarelli, A., Bollati, V., Tarantini, L., Moore, L. E., & Bonefeld-Jorgensen, E. C. (2008). Global DNA hypomethylation is associated with high serum-persistent organic pollutants in Greenlandic Inuit. *Environmental Health Perspectives, 116,* 1547–1552.

13. Yauk, C., Polyzos, A., Rowan-Carroll, A., Somers, C. M., Godschalk, R. W., Van Schooten, F. J., et al. (2008). Germ-line mutations, DNA damage, and global hyper-methylation in mice exposed to particulate air pollution in an urban/industrial location. *Proceedings of the National Academy of Sciences of the United States of America, 105,* 605–610.

14. Huang, Y.-T., VanderWeele, T. J., & Lin, X. (2014). Joint analysis of SNP and gene expression data in genetic association studies of complex diseases. *Annals of Applied Statistics, 8,* 352–376.

15. Huang, Y-T. (2015). Integrative modeling of multi-platform genomic data under the framework of mediation analysis. *Statistics in Medicine, 34,* 162–178.

16. Probst, A. V., Dunleavy, E., & Almouzni, G. (2009). Epigenetic inheritance during the cell cycle. *Nature Reviews. Molecular Cell Biology, 10,* 192–206.

17. Chan, T. L., Yuen, S. T., Kong, C. K., Chan, Y. W., Chan, A. S., Ng, W. F., et al. (2006). Heritable germline epimutation of MSH2 in a family with hereditary non-polyposis colorectal cancer. *Nature Genetics, 38,* 1178–1183.

18. Hitchins, M. P., Wong, J. J., Suthers, G., Suter, C. M., Martin, D. I., Hawkins, N. J., et al. (2007). Inheritance of a cancer-associated MLH1 germ-line epimutation. *New England Journal of Medicine, 356,* 697–705.

19. Suter, C. M., Martin, D. I., & Ward, R. L. (2004). Germline epimutation of MLH1 in individuals with multiple cancers. *Nature Genetics, 36,* 497–501.

20. Holliday, R. (1987). The inheritance of epigenetic defects. *Science, 238,* 163–170.

21. Rakyan, V. K., Blewitt, M. E., Druker, R., Preis, J. I., & Whitelaw, E. (2002). Metastable epialleles in mammals. *Trends in Genetics, 18,* 348–351.

22. Youngson, N. A., & Whitelaw, E. (2008). Transgenerational epigenetic effects. *Annual Review of Genomics and Human Genetics, 9,* 233–257.

23. Whitelaw, N. C., & Whitelaw, E. (2008). Transgenerational epigenetic inheritance in health and disease. *Current Opinion in Genetics & Development, 18,* 273–279.

24. Bollati, V., & Baccarelli, A. (2010). Environmental epigenetics. *Heredity, 105,* 105–112.

25. Morgan, D. K., & Whitelaw, E. (2008). The case for transgenerational epigenetic inheritance in humans. *Mammalian Genome, 19,* 394–397.

26. Jirtle, R. L., & Skinner, M. K. (2007). Environmental epigenomics and disease susceptibility. *Nature Reviews. Genetics, 8,* 253–262.

27. Rakyan, V., & Whitelaw, E. (2003). Transgenerational epigenetic inheritance. *Current Biology, 13,* R6.

28. Dolinoy, D. C. (2007). Epigenetic gene regulation: Early environmental exposures. *Pharmacogenomics, 8,* 5–10.

29. Cortessis, V. K., Thomas, D. C., Levine, A. J., Breton, C. V., Mack, T. M., Siegmund, K. D., et al. (2012). Environmental epigenetics: Prospects for studying epigenetic mediation of exposure-response relationships. *Human Genetics, 131,* 1565–1589.

30. Morgan, H. D., Sutherland, H. G., Martin, D. I., & Whitelaw, E. (1999). Epigenetic inheritance at the agouti locus in the mouse. *Nature Genetics, 23,* 314–318.

31. Waterland, R. A., & Jirtle, R. L. (2003). Transposable elements: Targets for early nutritional effects on epigenetic gene regulation. *Molecular and Cellular Biology, 23,* 5293–5300.

32. Dolinoy, D. C., Weidman, J. R., Waterland, R. A., & Jirtle, R. L. (2006). Maternal genistein alters coat color and protects A^{vy} mouse offspring from obesity by modifying the fetal epigenome. *Environmental Health Perspectives, 114,* 567–572.

33. Lillycrop, K. A., Phillips, E. S., Jackson, A. A., Hanson, M. A., & Burdge, G. C. (2005). Dietary protein restriction of pregnant rats induces and folic acid supplementation prevents epigenetic modification of hepatic gene expression in the offspring. *Journal of Nutrition, 135,* 1382–1386.

34. Cropley, J. E., Suter, C. M., Beckman, K. B., & Martin, D. I. (2006). Germ-line epigenetic modification of the murine A^{vy} allele by nutritional supplementation.

Proceedings of the National Academy of Sciences of the United States of America, 103, 17308–17312.

35. Anway, M. D., Cupp, A. S., Uzumcu, M., & Skinner, M. K. (2005). Epigenetic transgenerational actions of endocrine disruptors and male fertility. *Science, 308,* 1466–1469.

36. Dolinoy, D. C., Huang, D., & Jirtle, R. L. (2007). Maternal nutrient supplementation counteracts bisphenol A-induced DNA hypomethylation in early development. *Proceedings of the National Academy of Sciences of the United States of America, 104,* 13056–13061.

37. Pembrey, M. E., Bygren, L. O., Kaati, G., Edvinsson, S., Northstone, K., Sjostrom, M., et al. (2006). Sex-specific, male-line transgenerational responses in humans. *European Journal of Human Genetics, 14,* 159–166.

38. Lumey, L. H. (1992). Decreased birthweights in infants after maternal in utero exposure to the Dutch famine of 1944–1945. *Paediatric and Perinatal Epidemiology, 6,* 240–253.

39. Li, Y. F., Langholz, B., Salam, M. T., & Gilliland, F. D. (2005). Maternal and grandmaternal smoking patterns are associated with early childhood asthma. *Chest, 127,* 1232–1241.

40. Yauk, C. L., Berndt, M. L., Williams, A., Rowan-Carroll, A., Douglas, G. R., & Stampfli, M. R. (2007). Mainstream tobacco smoke causes paternal germ-line DNA mutation. *Cancer Research, 67,* 5103–5106.

41. Skinner, M. K. (2008). What is an epigenetic transgenerational phenotype? F3 or F2. *Reproductive Toxicology, 25,* 2–6.

42. Dolinoy, D. C., & Jirtle, R. L. (2008). Environmental epigenomics in human health and disease. *Environmental and Molecular Mutagenesis, 49,* 4–8.

43. Faucett, C. L., & Thomas, D. C. (1996). Simultaneously modelling censored survival data and repeatedly measured covariates: A Gibbs sampling approach. *Statistics in Medicine, 15,* 1663–1685.

44. Elashoff, R. M., Li, G., & Li, N. (2007). An approach to joint analysis of longitudinal measurements and competing risks failure time data. *Statistics in Medicine, 26,* 2813–2835.

45. Faucett, C. L., Schenker, N., & Elashoff, R. M. (1998). Analysis of censored survival data with intermittently observed time-dependent binary covariates. *Journal of the American Statistical Association, 93,* 427–437.

46. Li, N., Elashoff, R. M., & Li, G. (2009). Robust joint modeling of longitudinal measurements and competing risks failure time data. *Biometrical Journal/Biometrische Zeitschrift, 51,* 19–30.

47. Ibrahim, J. G., & Molenberghs, G. (2009). Missing data methods in longitudinal studies: a review. *Test (Madr)*, *18*, 1–43.

48. Deslandes, E., & Chevret, S. (2010). Joint modeling of multivariate longitudinal data and the dropout process in a competing risk setting: Application to ICU data. *BMC Medical Research Methodology*, *10*, 69.

49. Yuan, Y., & MacKinnon, D. P. (2009). Bayesian mediation analysis. *Psychological Methods*, *14*, 301–322.

50. MacKinnon, D. P., Lockwood, C. M., Hoffman, J. M., West, S. G., & Sheets, V. (2002). A comparison of methods to test mediation and other intervening variable effects. *Psychological Methods*, *7*, 83–104.

51. MacKinnon, D. P., Lockwood, C. M., Brown, C. H., Wang, W., & Hoffman, J. M. (2007). The intermediate endpoint effect in logistic and probit regression. *Clinical Trials*, *4*, 499–513.

52. Pearl, J. (2012). The Causal Mediation Formula: A guide to the assessment of pathways and mechanisms. *Prevention Science*, *13*, 426–436.

53. Imai, K., Keele, L., & Tingley, D. (2010). A general approach to causal mediation analysis. *Psychological Methods*, *15*, 309–334.

54. Robins, J. M., & Greenland, S. (1992). Identifiability and exchangeability for direct and indirect effects. *Epidemiology*, *3*, 143–155.

55. Breton, C. V., Salam, M. T., Wang, X., Byun, H. M., Siegmund, K. D., & Gilliland, F. D. (2012). Particulate matter, DNA methylation in nitric oxide synthase, and childhood respiratory disease. *Environmental Health Perspectives*, *120*, 1320–1326.

56. Islam, T., Breton, C., Salam, M. T., McConnell, R., Wenten, M., Gauderman, W.J., et al. (2010). Role of inducible nitric oxide synthase in asthma risk and lung function growth during adolescence. *Thorax*, *65*, 139–145.

57. Salam, M. T., Byun, H. M., Lurmann, F., Breton, C. V., Wang, X., Eckel, S. P., & Gilliland, F. D. Genetic and epigenetic variations in inducible nitric oxide synthase promoter, particulate pollution, and exhaled nitric oxide levels in children. *Journal of Allergy and Clinical Immunology* 2012; 129: 232–9 e1–7

58. Breton, C. V., Byun, H.-M., Wang, X., Salam, M. T., Siegmund, K., & Gilliland, F. D. (2011). DNA methylation in the arginase-nitric oxide synthase pathway is associated with exhaled nitric oxide in children with asthma. *American Journal of Respiratory and Critical Care Medicine*, *184*, 191–197.

59. Benjamin, R., Kathleen, B., Terri, H. B., Anthony, B., Vincent, J. C., Mario, C., et al. Asthma Bridge: The asthma biorepository for integrative genomic exploration. In *D101: Asthma Genetics*, American Thoracic Society No. pp. A6189–A6189.

60. Siegmund, K. D., & Laird, P. W. (2002). Analysis of complex methylation data. *Methods (San Diego, Calif.)*, *27*, 170–178.

61. Siegmund, K. D., Laird, P. W., & Laird-Offringa, I. A. (2004). A comparison of cluster analysis methods using DNA methylation data. *Bioinformatics*, *20*, 1896–1904.

62. Siegmund, K. D., Levine, A. J., Chang, J., & Laird, P. W. (2006). Modeling exposures for DNA methylation profiles. *Cancer Epidemiology, Biomarkers & Prevention*, *15*, 567–572.

63. Bottolo, L., Chadeau-Hyam, M., Hastie, D. I., Zeller, T., Liquet, B., Newcombe, P., et al. (2013). GUESS-ing polygenic associations with multiple phenotypes using a GPU-based evolutionary stochastic search algorithm. *PLOS Genetics*, *9*, e1003657.

64. Papathomas, M., Molitor, J., Hoggart, C., Hastie, D., & Richardson, S. (2012). Exploring data from genetic association studies using Bayesian variable selection and the Dirichlet process: Application to searching for gene × gene patterns. *Genetic Epidemiology*, *36*, 663–674.

65. Molitor, J., Marjoram, P., & Thomas, D. (2003). Application of Bayesian spatial statistical methods to analysis of haplotype effects and gene mapping. *Genetic Epidemiology*, *25*, 95–105.

66. Molitor, J., Marjoram, P., & Thomas, D. C. (2003). Fine-scale mapping of diseases with multiple mutations via spatial clustering techniques. *American Journal of Human Genetics*, *73*, 1368–1384.

67. Thomas, D. C., Stram, D. O., Conti, D., Molitor, J., & Marjoram, P. (2003). Bayesian spatial modeling of haplotype associations. *Human Heredity*, *56*, 32–40.

68. Thomas, D. C. (2007). Multistage sampling for latent variable models. *Lifetime Data Analysis*, *13*, 565–581.

69. Thomas, D. C., Conti, D. V., Baurley, J., Nijhout, F., Reed, M., & Ulrich, C. M. (2009). Use of pathway information in molecular epidemiology. *Human Genomics*, *4*, 21–42.

70. van den Berg, B. J., Christianson, R. E., & Oechsli, F. W. (1988). The California Child Health and Development Studies of the School of Public Health, University of California at Berkeley. *Paediatric and Perinatal Epidemiology*, *2*, 265–282.

71. Cohn, B. A., Cirillo, P. M., & Christianson, R. E. (2010). Prenatal DDT exposure and testicular cancer: A nested case-control study. *Archives of Environmental & Occupational Health*, *65*, 127–134.

72. Cohn, B. A., Cirillo, P. M., Sholtz, R. I., Ferrara, A., Park, J. S., & Schwingl, P. J. (2011). Polychlorinated biphenyl (PCB) exposure in mothers and time to pregnancy in daughters. *Reproductive Toxicology*, *31*, 290–296.

73. Cohn, B. A., Cirillo, P. M., Wolff, M. S., Schwingl, P. J., Cohen, R. D., Sholtz, R. I., et al. (2003). DDT and DDE exposure in mothers and time to pregnancy in daughters. *Lancet, 361,* 2205–2206.

74. Jaquish, C. E. (2007). The Framingham Heart Study, on its way to becoming the gold standard for cardiovascular genetic epidemiology? *BMC Medical Genetics, 8,* 63.

75. Cupples, L. A., Arruda, H. T., Benjamin, E. J., D'Agostino, R. B., Sr., Demissie, S., DeStefano, A. L., et al. (2007). The Framingham Heart Study 100K SNP genome-wide association study resource: Overview of 17 phenotype working group reports. *BMC Medical Genetics, 8*(Suppl 1), S1.

76. North, K. E., Goring, H. H., Cole, S. A., Diego, V. P., Almasy, L., Laston, S., et al. (2006). Linkage analysis of LDL cholesterol in American Indian populations: The Strong Heart Family Study. *Journal of Lipid Research, 47,* 59–66.

77. North, K. E., Howard, B. V., Welty, T. K., Best, L. G., Lee, E. T., Yeh, J. L., et al. (2003). Genetic and environmental contributions to cardiovascular disease risk in American Indians: The Strong Heart Family Study. *American Journal of Epidemiology, 157,* 303–314.

78. Troisi, R., Hyer, M., Hatch, E. E., Titus-Ernstoff, L., Palmer, J. R., Strohsnitter, W. C., et al. (2013). Medical conditions among adult offspring prenatally exposed to diethylstilbestrol. *Epidemiology, 24,* 430–438. doi:.10.1097/EDE.0b013e318289bdf7

79. Titus-Ernstoff, L., Troisi, R., Hatch, E. E., Hyer, M., Wise, L. A., Palmer, J. R., et al. (2008). Offspring of women exposed in utero to diethylstilbestrol (DES): A preliminary report of benign and malignant pathology in the third generation. *Epidemiology, 19,* 251–257. doi:.10.1097/EDE.0b013e318163152a

Contributors

Fatima Umber Ahmed, California National Primate Research Center

Yin-Hsiu Chen, University of Michigan, Ann Arbor

James Y. Dai, Fred Hutchinson Cancer Research Center, Seattle

Caroline Y. Doyle, Alpert School of Medicine at Brown University, Providence

Zihuai He, University of Michigan, Ann Arbor

Li Hsu, Fred Hutchinson Cancer Research Center, Seattle

Shuo Jiao, Fred Hutchinson Cancer Research Center, Seattle

Erin Loraine Kinnally, University of California-Davis, California National Primate Research Center

Yi-An Ko, University of Michigan, Ann Arbor

Charles Kooperberg, Fred Hutchinson Cancer Research Center, Seattle

Seunggeun Lee, University of Michigan, Ann Arbor

Arnab Maity, North Carolina State University, Raleigh

Jeanne M. McCaffery, Alpert School of Medicine at Brown University, Providence

Bhramar Mukherjee, University of Michigan, Ann Arbor

Sung Kyun Park, University of Michigan, Ann Arbor

Duncan C. Thomas, University of Southern California, Los Angeles

Alexandre Todorov, Washington University, St. Louis

Jung-Ying Tzeng, North Carolina State University, Raleigh

Tao Wang, Albert Einstein College of Medicine, New York

Michael Windle, Emory University, Atlanta

Min Zhang, University of Michigan, Ann Arbor

Index